The Role of Physics Departments in Preparing K–12 Teachers

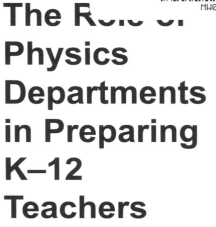

Edited by:
Gayle A. Buck
Jack G. Hehn
Diandra L. Leslie-Pelecky

Sponsored by:
The University of Nebraska–Lincoln
The American Institute of Physics
The American Physical Society
The American Association of
Physics Teachers
Nebraska EPSCoR

The following papers were prepared for the working conference, "The Role of Physics Departments in Preparing K–12 Teachers" at the University of Nebraska–Lincoln in June 2000. Slight revisions for publication have been made to each paper. The views expressed do not necessarily represent those of the conference sponsors.

The American Institute of Physics (AIP) was founded in 1931 and chartered as a membership corporation for the advancement and diffusion of knowledge of the science of physics and its application to human welfare.

To achieve its purpose, AIP serves the sciences of physics and astronomy by serving the Member Societies, individual scientists, students, and the general public.

Copyright © 2000 by The American Institute of Physics.
All rights reserved.
ISBN# 1-56396-985-8

Cover
The cover image depicts the ideal relationship between teacher and learner. The teacher offers the support and foundation to the learner, who, in turn, takes an active role and reaches to discover new knowledge.

Web site of The Role of Physics Departments in Preparing Future Teachers: www.physics.unl.edu/DLP/TeachersConference.html

For more information contact:

American Institute of Physics
Education
One Physics Ellipse
College Park, MD 20740-9910
Tel: 301-209-3007
Fax: 301-209-0839
Webpage: www.aip.org

# The Role of Physics Departments in Preparing K–12 Teachers

A Working Conference

June 8 & 9, 2000
at the University of Nebraska–Lincoln

The Challenge

Improving the scientific preparation of prospective K–12 teachers has received dramatically increased attention and support in the last eight years. State and regional accountability efforts have included the adoption of state science standards, often based on a small number of national models. These standards mandate science content knowledge, a thorough understanding of the process and context of science, and familiarity with technology as a tool for learning. States and local school systems are changing accreditation and hiring requirements in response to the new standards. There is a perception that colleges and universities are not adapting rapidly enough to prepare new teachers to meet the challenges that they will face.

Teacher preparation has been identified as a federal priority in budget efforts of both Congress and the Executive Branch. The predicted need for two million new teachers within the next decade will strain an already burdened system of teacher preparation. Professional societies of mathematicians and scientists have supported statements that encourage discipline-based departments to engage more vigorously and collaboratively in the process of preparing future teachers, recognizing that all elementary school teachers are teachers of science and mathematics.

Conference Goals

To evaluate the range of current pedagogical approaches and the evidence for the efficacy of these approaches to teaching science to preservice teachers.

To investigate different implementations of new programs for teacher education with a focus on understanding the elements necessary for implementing, sustaining, and institutionalizing such a program.

To compile the information obtained from this study by soliciting papers from leaders in the field that address the key elements of the items listed above and to share this information with institutions with similar goals.

Statement on the Education of Future Teachers

The scientific societies listed below urge the physics community, specifically physical science and engineering departments and their faculty members, to take an active role in improving the preservice training of K–12 physics and science teachers. Improving teacher training involves building cooperative working relationships between physicists in universities and colleges and the individuals and groups involved in teaching physics to K–12 students. Strengthening the science education of future teachers addresses the pressing national need for improving K–12 physics education and recognizes that these teachers play a critical education role as the first, and last, physics teacher for most students.

While this responsibility can be manifested in many ways, research indicates that effective preservice education involves hands-on, laboratory-based learning. Good science and mathematics education will help create a scientifically literate public, capable of making informed decisions on public policy involving scientific matters. A strong K–12 physics education is also the first step in producing the next generation of researchers, innovators, and technical workers.

American Institute of Physics
American Physical Society
American Association of Physics Teachers
American Astronomical Society
Acoustical Society of America
American Association of Physicists in Medicine
American Vacuum Society

Table of Contents

Context: Why Have a Conference on Teacher Preparation?...... 1
Diandra L. Leslie-Pelecky, University of Nebraska–Lincoln
Gayle A. Buck, University of Nebraska–Lincoln

Supporting Papers

How Can a Physics Course for Non-majors and Preservice K–8 Teachers Engage Students in the Process of Scientific Inquiry?... 7
Lynn M. Tashiro, California State University, Sacramento

Curricula Gridlock and Other Challenges in Teacher Preparation.. 43
Karen L. Johnston, North Carolina State University

How Computer Technology Can Be Incorporated into a Physics Course for Prospective Elementary Teachers........ 53
Fred Goldberg, San Diego State University

Preparing Teachers to Teach Physics and Physical Science by Inquiry.. 71
Lillian C. McDermott, University of Washington
Peter S. Shaffer, University of Washington

Investigating the Role of Physics Departments in the Preparation of K–12 Teachers 87
John Layman, University of Maryland

The Role of Physics Departments in Preservice Teacher Preparation: Obstacles and Opportunities........................ 109
José Mestre, University of Massachusetts

Re-Preparing the Secondary Teacher............................... 131
Dr. Fredrick M. Stein, American Physical Society

Keynote Address: L. Dennis Smith, University of Nebraska..... 141

Acknowledgements.. 145

Context

WHY HAVE A CONFERENCE ON TEACHER PREPARATION?

Improving the scientific preparation of prospective K–12 teachers has received dramatically increased attention and support in recent years. State and regional accountability efforts have included the adoption of state science standards, often based on a small number of national models. These standards mandate science content knowledge, a thorough understanding of the process and context of science, and familiarity with technology as a tool for learning. States and local school systems are changing accreditation and hiring requirements in response to the new standards. There is a perception that colleges and universities are not adapting rapidly enough to prepare new teachers to meet the challenges that they will face.

Teacher preparation has been identified as a federal priority in budget efforts of both Congress and the Executive Branch. The predicted need for two million new teachers within the next decade strain an already burdened system of teacher preparation. Professional societies of mathematicians and scientists have supported statements that encourage discipline-based departments to engage more vigorously and collaboratively in the process of preparing future teachers, recognizing that all elementary school teachers are teachers of science and mathematics.

Statements of principle made by professional societies, legislators, and funding agencies are one thing, but advances in education ultimately start with "in-the-trenches" educators trying to address a specific problem on the local scale. A university-wide initiative on math and science education at the University of Nebraska–Lincoln brought together interested faculty from math, science, engineering, and education. These working groups showed us that, while we were all in agreement that improvements were necessary, there were, between educators and physicists, rather large cultural (and spatial) gaps, that would have to be bridged if reform efforts were to be truly meaningful.

Assistant professors tend to have a finite time horizon, so while campus-wide plans were being made for an infrastructure in support of math and science education, we opted to focus our efforts on developing a one-semester course for prospective K–8 teachers. We thought that this would be an easier, more tractable problem; however, the more we investigated, the more we realized that even this small component has endless complexities. Research shows that traditional college physics courses are probably not providing students with the skills they need to learn enough science to be able to teach. At the same time, temporal and financial constraints cannot be ignored when developing a new class. Add to this state- and federal-mandated accountability standards and the problem begins to look insurmountable.

We began our research with a literature survey to see what had already been done or developed. The literature shows a broad range of approaches that are documented in varying levels of detail. Surveying conference abstracts showed a lot of valuable work that never makes it to the formal literature. We found ourselves raising a lot of questions that weren't answered in the literature.

An advisor in graduate school suggested that the best way to get into a new field is to read a review article—or write one. What we felt was missing from our literature survey was a sense of interaction—of having the opportunity to have people with experience in these fields talk to each other and to us about why they made the choices they did. We decided that the most efficient way to gather information would be to obtain funding from the university and hold a "mini-conference." We originally planned to bring three or four experts to campus for a one-day intensive meeting that would provide us with the chance to "compare and contrast" different approaches.

We soon realized that our concerns about preparing teachers were shared by many other people. The professional physics societies had recently endorsed a statement of the importance of physics departments becoming involved in teacher preparation. A discussion at the American Center for Physics led to a collaboration with the American Institute of Physics (AIP), the American Physical Society (APS) and the American Association of Physics Teachers (AAPT) that allowed us to increase the number of speakers, broaden the focus from K–8 to K–12, open the conference to people outside the university, and produce a portfolio of resources that would address some of the questions that we—and many other educators—were asking about teacher preparation.

"The Role of Physics Departments in Preparing K–12 Teachers" was held on June 9, 2000 at the University of Nebraska–Lincoln. We could not possibly address all of the relevant issues in a one-day conference, so we had to make some difficult up-front choices to prevent the conference from becoming an incoherent race through every possible topic. Our selections of speakers and topics were heavily influenced by our particular situation. From a long list of physicists and educators, we chose speakers who were physicists at research-intensive public universities. A number of very deserving speakers had to be omitted due to lack of time. We selected a representative range of approaches that could be compared with each other and serve as the foundation for further exploration.

Two questions struck us as fundamental. First, what pedagogical approaches exist and what data demonstrate the efficacy of these approaches with preservice teachers? Second, how can a new course—especially one that may be personnel intensive—be initiated and sustained within departmental resource constraints? Our speakers were thus charged with the difficult task of addressing a broad range of concerns. We think they did an admirable job, both during the conference, and in their written papers.

An important part of this conference was our desire to avoid making it seven consecutive colloquia. We wanted to have conversations between speakers, and between the speakers and the conference participants. We thus allowed the speakers a very short time—twenty minutes—for their oral presentations. The remainder of the time was organized around panels that focused on eight questions. Each topic was briefly discussed by small groups of participants at their tables and each table submitted three questions to be asked of a panel formed of a subset of the speakers. Further discussion between the speakers, and contributions from the conference participants, made this a very effective mode of operation. The most frustrating part of this activity was having to cut off interesting discussions when it was time to move to the next topic!

We are very thankful to the conference participants—especially the invited speakers who put a lot of effort into their presentations and their papers. A special recognition must also be extended to The University of Nebraska–Lincoln's President, L. Dennis Smith, whose luncheon address set the effort in a national context (see pages 141-144). We learned an incredible amount from the conference and our interactions with the speakers and other participants. We found this to be a very efficient mode for educating ourselves; however, we also realized that there are many issues that we were either unable to address or that we had to address in unsatisfactory brevity.

Overall, we must agree with the findings of the conference evaluator, Karen L. Johnston from the Momentum Group of Fort Worth, Texas. She reported, in part,

> *"The conference on the role of physics departments in teacher preparation achieved its objectives. Discussions, both formal and informal, among physics and education faculty reveal a readiness to begin the dialog and prepare for the collaborative work associated with better preparation of K–12 teachers. Physics professional societies play a pivotal role in changing expectations for physics departments regarding their responsibilities in teacher preparation. Faculty, departments, institutions and professional societies all have roles in this important endeavor."*

One of the most important aspects of evaluating a conference is raising the inevitable question, "if you had it to do over again, what would you do differently?" Given the importance of the collaboration between faculty in Teachers' College and in the Department of Physics & Astronomy in our particular initiative, we would have liked to have had time to present more of the perspective of working teachers, teachers-to-be, and education faculty. We also would have liked more than one day to discuss these very important but complex issues. We are very happy with the results of this initial effort and hope that you find the collected resources as useful as we do.

Thanks are due to a number of people for their support of our efforts to make this conference happen. As assistant professors, we could not have proceeded without the unwavering support of our department chairs, Roger Kirby (Physics & Astronomy) and Beth Franklin (Curriculum & Instruction). The Math/Science Education Initiative, directed by Jim Lewis, and the Center for Math, Science & Computer Education, directed by Sandra Scofield are responsible for establishing an environment on campus that promotes discussion of these issues across departmental and college boundaries. We are grateful to Brian Foster, James O'Hanlon, Linda Pratt, and Laura White for their foresight in establishing the infrastructure that makes activities like ours possible. We acknowledge financial support from Royce Ballinger (Nebraska EPSCoR), the College of Arts & Sciences, Teachers' College, and the Center for Math, Science & Computer Education, as well as the administrative support from our respective departments.

Finally, we must thank the friends and associates who made wonderful suggestions and provided sympathetic ears. Many thanks to Patrick Buck, Robert C. Hilborn, Suzanne Kirby, and Vicki Plano Clark. A special thanks to Marilyn McDowell for all of her invaluable assistance with planning the logistics of the conference.

Diandra L. Leslie-Pelecky & Gayle A. Buck, Conference Co-Chairs

SUPPORTING PAPERS

How Can a Physics Course for Non-majors and Preservice K–8 Teachers Engage Students in the Process of Scientific Inquiry? A Case Study in Collaborative Curriculum Design and Implementation

Lynn M. Tashiro
California State University Sacramento

This case study documents a collaborative process of designing a physics course across the academic cultures of science and education. It presents a model for how scientific inquiry might be integrated into an undergraduate physics course for non-majors and preservice K–8 teachers. Tools for managing and assessing guided and open inquiry are provided together with samples of student work. Evidence of student learning is investigated, with a focus on assessing students' ability to write a testable scientific question.

The paper is organized in five sections:

I. The context and resources for designing the physics course, Physics 100
II. The collaborators
III. The course
IV. Tools for managing and assessing inquiry
V. Evidence of student learning
VI. References

I. The Context: What was the Motivation for Designing a Physics Course for Preservice K–8 Teachers? What Resources Did We Have?

The design of Physics 100 was motivated by a desire to prepare our future teachers to meet the challenges of implementing national and state science standards. Of particular concern was developing a curriculum that would provide experience with and develop skills necessary to engage in scientific inquiry. In the fall of 1998 the National Science Foundation awarded funding to California State University Sacramento (CSUS) for project C-CUESST (A College

Curriculum for Elementary School Science Teachers). The goal of the project was to create an inquiry-based undergraduate physics course, Physics 100, for preservice K–8 teachers that would integrate scientific content knowledge with research-based knowledge about teaching and learning science. Faculty release time for the collaborators and the purchase of classroom computers were funded by NSF ($247K) and CSUS ($15K). This project was the first step in redesigning the entire preservice science curriculum at CSUS.

II. The Collaborators: Who Designed and Taught Physics 100?

Lynn Tashiro
Associate Professor
of Physics

Steve Gregorich
Professor of
Teacher
Education

Patricia McEgan
K–8 science teacher

Hugo Chacón
Assistant Professor
of Bilingual and
Multicultural Education

Lynn, Steve, Patty and Hugo are the collaborative team that planned and team taught Physics 100. Formal project evaluation is being conducted by David Jelinek, CSUS Assistant Professor of Science Education.

How Did We Design and Implement the Course and Who Did What?

The following timeline will help to understand the context of each team member's work:

Course design and implementation timeline

- Lynn and Steve planned and put together materials for the guided inquiry lessons
- Search conducted for the third collaborator, the K–8 teacher in residence

Spring 1999 semester
- Patty joined the project full time as the teacher in residence

- Steve and Lynn team taught two pilot sections of Physics 100.
- Guided inquiry and open inquiry projects were tested
- Guided and open inquiry components modified as a result of formative evaluation feedback from students and project external evaluator

Fall 1999 semester
- Lynn and Patty team taught two sections of Physics 100
- Transitioned four additional sections of Physics 100 and part-time faculty to the inquiry-based curriculum
- Hugo joined the project
- Hugo and Steve began revision of the science methods course
- Hugo and Lynn redesigned and team taught the light and color unit

Spring 2000 semester
- Lynn taught two sections of Physics 100
- Patty and Hugo each taught a science methods course
- A limited enrollment learning community consisting of Physics 100, EDTE306 Science Methods and EDBM470 a field experience was piloted by Lynn and Hugo
- Patty began work on a master's thesis, which will use discourse analysis to investigate students' experience and perception of the open inquiry process
- Together with Professor Melanie Loo of biology, Lynn piloted an early field experience in teaching K–6 science for preservice teachers, ID111 Science in the Elementary School
- David, Lynn and Patty began evaluating students' ability to ask a testable question.

The contributions of each team member were based not only on their areas of academic expertise but also as a result of individual interests and talents. Key contributions of team members are summarized below:

Lynn (physics content specialist)
- provided the science content and activities for the guided inquiry lessons ensuring a level of rigor acceptable to university physics and science faculty
- reviewed and presented literature on physics and K–12 science education
- created the structure and management tools for the open inquiry projects and poster presentations
- wrote learning objectives, the "know and do boxes," for each guided inquiry lesson
- evaluated students' ability to ask testable question

Steve (science methods specialist)
- organized, formatted and wrote the first draft of the student laboratory manual containing the guided inquiry activities
- researched, selected and constructed instructional materials, measurement tools and technology components in the sound and electricity units.
- operationalized the guided inquiry learning objectives, the "know and do boxes," by focusing them on behavioral objectives.
- provided a global perspective of how our course fits into the teacher education program. As past dean of the School of education, Steve provided invaluable insight and help in obtaining administrative support for our project

Patricia (K–8 science teaching specialist)
- constructed the first draft of the open inquiry rubric used to assess the open inquiry projects. (It was modeled after a rubric she uses in her K–8 science classroom.)
- edited the laboratory manual improving it by the addition of focusing question for each activity.
- provided reality check on the content and process skill objectives in Physics 100, making sure activities and projects were relevant to state and national standards as well as science in the K–8 classroom

Hugo (science education and multicultural education specialist)
- designed guided activities in the light and color unit
- expanded our learning assessment plan by using oral presentations as the culminating open inquiry activity
- created a link between science methods, science in the K–8 classroom and Physics 100 by offering a Learning Community together with Lynn consisting of Physics 100, EDTE306 (Science Curriculum and Instruction) and EDBM470 (Teaching Science in the Elementary School)

Over the past three semesters all four of us have team taught some portion of Physics 100 together and participated in facilitating and evaluating the open inquiry projects.

Some important factors in facilitating collaborative work:
- **Matching personalities and teaching styles**
  Matching personalities and teaching styles is just as important as matching academic expertise in building a collaborative team. Project C-CUESST has a very precious collaborative component. All team members were flexible, constructively critical and willing to try just about anything to improve the learning in Physics 100. The team members shared enough common

knowledge and teaching philosophy to have productive discussions but were diverse enough in their thinking so that multiple viewpoints and solutions to problems were always presented.

- **Strong physics and pedagogy content knowledge by each collaborator**
Each of the team members had a strong background in physical science and science education research. Steve was the science and technology instructor for the school of education and had been involved in the move toward inquiry in the 1970s. Patricia had a strong physical science background and was involved in the California systemic projects to improve K–8 science teaching. Hugo had been a high school physics teacher as well as a university science methods instructor, and Lynn, in addition to teaching undergraduate physics courses, has been involved with science education research and state systemic projects to improve K–8 science learning.
- **Providing physical space and opportunity for collaboration**
Physically sharing office space and participating in each other's department meetings and social functions is important. At the beginning of the project Steve, Patty and Lynn moved into an office in the science building where Physics 100 is taught. During the project Steve, Patty and Hugo attended several physics department meetings and social events. Lynn attended meetings with CSUS education faculty and K–6 teachers within the school of education as well as at K–6 school sites in each of the local area districts. These small events helped each of us to observe the differences, understand the constraints and identify the opportunities present in the academic cultures of the university science departments, university education departments and K–6 classrooms.

Out of this collaboration emerged a practical model of an inquiry-based physics course. This model is described in detail in the following section.

III. The Course: What does Physics 100 Look Like?

How is Physics 100 Organized?

Physics 100 meets for an hour and 40 minutes twice a week (200 minutes of "activity time" per week) in a laboratory style classroom designed to seat 24 students.

Topically the course is divided into three units, waves and sound, electricity and magnetism, and light and color. These topics were chosen to fit into the existing preservice science curriculum. Physics 100 is one of seven science courses required of preservice teachers and is the second of two physics courses required.

Instructionally Physics 100 is composed of a lecture/textbook component, a guided inquiry component and an open inquiry component. The approximate distribution of class time allocated for each component is shown below:

Distribution of instructional modes

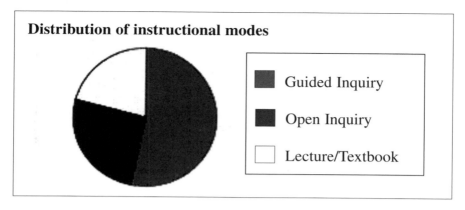

What Do the Components Look Like?
A description of each component follows:

The Lecture/textbook component:
The lecture/textbook component is modeled after instruction that is traditional in the physics discipline. Weekly textbook reading from Hewett's *Conceptual Physics* is assigned along with end of chapter review questions, discussion questions and numerical problems. Lectures are integrated into the guided inquiry component and are used to introduce, formalize, or summarize concepts. The lectures are more like "minilectures" and are usually 10 to 20 minutes in duration and focus on a physicist's understanding of concepts. For example a minilecture is used to formalize, define and differentiate the electrical concepts of voltage, current, resistance and power.

The Guided Inquiry component:
The guided inquiry component is built on theoretical foundations of physics education research[1-5] and K–8 science education research.[6,7] The guided inquiry lessons have four phases we call "into," "through," "beyond" and "reflection."
The "into," "through," and "beyond" phases are similar to the phases in the learning cycle used with children[6] and the SCALE (Science Content And Language Expansion) cycle proposed as effective with culturally diverse children.[7] These three phases also include the four stages of the "modeling theory" 1. description, 2. formulation, 3. ramification and 4. validation presented by.[4]
The "into," "through," and "beyond" phases comprise the bulk of the guided inquiry lessons. The "reflection" phase can be thought of as the "student postmortem" of the guided activity. Although this reflection or metacognition exercise was identified in physics instruction as the period in the lesson where "the most significant learning occurs,"[4,5] it proved to be the most difficult to integrate and make room for in the time allocated for guided inquiry instruction. Although there are a few formally written examples of "reflection" lessons in the guided inquiry lessons, most of the metacognition happens informally at the end of each lesson

when the instructor and students discuss the "know and do boxes," which articulate what students should know and be able to do at the end of the guided inquiry lesson. More formal metacognition and thinking about learning lessons are being integrated into an experimental science methods course. This experimental course is offered in a learning community where students concurrently enroll in Physics 100, EDTE306 (Science Methods) and EDBM 470 (Teaching Science in an Elementary School).

For reference the "into," "through," and "beyond" phases are briefly described:

"into"
- probing for prior knowledge that requires students to expose their preconceptions and misconceptions about a concept or physical system
- determining relevant physical quantities to measure and include in the construction of a model of a system
- generating testable questions for a model and predicting outcomes of experiments based on a model

This stage may be initiated by posing a question or observation of a discrepant event (an event whose outcome is unexpected or counterintuitive)

"through"
- construction of a model for understanding observed phenomena
- designing experiments to answer the questions posed
- interpreting, debating, and defending experimental results
- validating, or debunking a model or prediction
- reflecting on the consequence of their experimental results
- presentation of results to peers

"beyond"
- deploying model to explain or predict the behavior of a related but new physical system
- applying a new skill or concept to another scientific discipline or personal event outside the laboratory

"reflection"
- students judging whether the goals articulated at the beginning of the lesson were achieved
- students reflecting upon and articulating which types of activities contributed most, and least to their learning

Some of the guided inquiry lessons were created by the collaborative team and some were modified from published texts such as *Physics by Inquiry* and Elementary Science Kits such as Science and Technology for Children (STC), and Full Option Science Systems (FOSS). Below are a few pages of the student laboratory workbook illustrating the "into," "through" and "beyond" components.

1: Wave Motion
Everything we see and hear reaches us by waves.

Into
Activity A. What is your past experience with waves? Use drawings, words, and numbers to illustrate what you already know about waves and what you do not understand about waves. 1. What are some examples of waves?

Through
Activity B. Can we observe waves traveling through different media? Three stations are set up around the room containing materials that can be made to produce waves. Below are pictures of the materials used for the stations. The questions for you to answer are on the following pages. Station 1: Pushing against one end of a slinky hung by strings from the celiling. 

Beyond
Activity F. Assessment: Construct and Observe a Wave Machine. Obtain a package of materials from your instructor and follow directions to construct a wave machine. (See directions for constructing the Wave Motion Machine in the Materials for Sound section at the end of this unit.) Observe waves that are created when you pull the fishing line taut and very gently and slowly twist your wrist just a little left and right, or  hold the machine very still and lightly tap one end of the soda straw nearst you. 1: How is the wave created?

2. What do you already know about waves?   Page 5

Station 2: Drops striking the surface of a puddle (milk carton & cake pan of water.

2. Identify the medium the wave is traveling through.

3. What direction is the medium vibrating?

4. What direction is the energy traveling?

5. What direction is the wave traveling?

6. Determine if the wave is longitudinal or transverse. Explain your answer using a diagram and words.

Page 10

Station 3: Making waves with a jump rope.

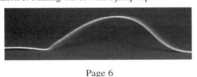

Page 6

The open inquiry component:

The open inquiry component was based on a model developed at the Institute for Inquiry at the San Francisco Exploratorium Museum and is described in "Volume 2 FOUNDATIONS, Inquiry: Thoughts, Views, and Strategies for the K–5 classroom."[8]

During the 15-week semester, students work in groups of four or five on three inquiry projects, one in each of the topic areas. Each project is assigned after students have had some concrete experience with phenomena in the topic area and is completed in a cycle of six class meetings. Each project begins with a student question that is then investigated by the group and finally presented in a scientific poster session. A timeline of an open inquiry cycle is shown below:

Timeline for student inquiry projects:

Day 1	Students are given the assignment of writing a testable question and investigation plan.
Day 2	Students discuss questions in groups and decide on one question to investigate
Day 3	Students are given 30 min of class time to work on inquiry
Day 4	Students are given 30 min of class time to work on inquiry
Day 5	Students are given 30 min of class time to work on inquiry and prepare poster
Day 6	One hour of class time is used for scientific poster session and peer evaluation of projects

An example of the open inquiry process for electricity and magnetism is illustrated below with the details of one of the six groups' experience described.

Day 1 Students are given the assignment of writing a testable question.

• Students are given an Inquiry Planning Sheet and Project Grading Criteria and asked to write a testable question and plan an investigation to answer the question.	**Inquiry Project Planning sheet** Name(s): Inquiry Question:
• They have had 2 and a half weeks of textbook, lecture and guided activity instruction on static electricity and electrical circuits.	What will you measure and what will you measure it with? What will you graph?
• The guided activities have instructed students on the use of the multimeter.	List of things you will provide for the inquiry: List of classroom materials you would like to use. (You may use any materials you have seen in class provided they are returned at the end of the class period. If you need something special consult your instructor):

Day 2 Students discuss questions in groups and decide on one question to investigate

• Each student is given two uninterrupted minutes to state and explain their question.

• Students then discuss the testability of the questions and choose one to pursue.

• If the group "gets stuck" the instructor facilitates the discussion

• When the group has come to consensus, they draft a group project planning sheet and review it with the instructor.

The questions this group is discussing are:

> 1. **How does the type of citrus used in a homemade battery (i.e. lemon battery) affect the voltage? Diana**
> 2. **How much do different diameters of wire affect the amount of charge through the wire? Andrea**
> 3. **Do different fruits provide different amounts of power to light a bulb? Neng**
> 4. **Which fruit or vegetable has the highest voltage or current? What causes this? Jenn**

During the conversation, the concepts of acidity and pH have come up as factors that might affect the voltage of their "fruit batteries" and the question the group agrees on is:

> **Does the pH level in different fruits affect their voltages?**

For the next inquiry meeting the group plans to bring fruits, zinc and copper screws, and pH paper. They have also sketched out their experimental setup.

Days 3, 4 and 5 Students are given 30 min of class time to work on inquiry and prepare posters

Students use a multimeter to measure the voltage produced by inserting zinc and copper screws into a kiwi. The multimeter measures 0.82 V.

Andrea and Jenn match the color of the pH paper to the chart to determine the pH of one of the fruits.

Dianna, Andrea, Neng and Jenn spend their in-class time using a multimeter to measure the voltage produced by various fruits as a result of inserting copper and zinc electrodes in them. They also obtain a pH measuring kit from a local school supply store and test the pH level of each of their fruits.

In between class meeting times this group has been surfing the Internet for background research to help them make sense of the data they are collecting. They have used the interactive inquiry journal to ask the instructor questions and respond to instructor feedback on their inquiry project.

Day 6 One hour of class time for scientific poster session and peer evaluation of projects

On the poster presentation day groups are busy assembling their posters. Dianna, Andrea, Neng, and Jenn pose with their finished product.

Students explain their posters to their peers and instructors.

Students are given about 10 minutes to assemble their posters. Some will have them completed before coming to class. Posters are set at the ends of the long lab tables in class and each inquiry group divides in half. Half of the group will stay by their poster and half will move to evaluate another poster. After 20 minutes the students switch so that they all have a chance to explain and look at other posters. Students are given a peer evaluation form for the posters that is the same as what the instructor will use to evaluate the posters.

On this day Professors David Jenlinek and Elizabeth Kean from the school of education and Professor Joong Lee from the Department of Chemistry have come to look at the inquiry posters. Students have engaged Professor Lee in a discussion of the chemical reaction between their hot dog and the zinc coated paper clips they used as electrodes in their inquiry. " Just what was that stuff oozing out of the hot dog?"

After the poster session, which lasts about an hour, students turn in a peer evaluation of one of the posters they examined. Students do not determine the grade of their peers but instead a student's grade is determined by their ability to apply the open inquiry rubric. Students are graded on their ability to provide evidence to support the grade they award a poster in four categories: appearance and organization, graphs and charts, analysis, and conclusion. The grade for the inquiry project will be based on the poster, the inquiry journal, attendance and their peer evaluations.

IV. Tools for Managing and Assessing Inquiry: What Tools Did We Find Useful and Necessary to Manage the Guided and Open Inquiry Components?

Guided Inquiry Management Tools: The Student Laboratory Workbook and "Know and Do Boxes".

The student laboratory workbook and the "know and do boxes" were useful for managing the guided inquiry process. The 135-page student workbook was an absolute necessity to make the course transferable to other instructors teaching the course who had not participated in its design. The workbook gave a definite structure to the course, but left enough room to accommodate different teaching styles. Samples of workbook pages are illustrated in section II of this paper.

The "know and do boxes" articulate what students should know and be able to do at the conclusion of the guided inquiry lesson. Part of the first "know and do box" for waves is shown below:

Physics 100
Concepts in Physics
Know and do box #1

What students should know:
1. The definitions of wave characteristics: l(wavelength), v (speed) f(frequency), T(period) and A (amplitude).
2. Scientific models can be drawings, graphs, computer simulations, mathematical equations or physical manipulatives.
3. Different models illustrate different characteristics of a physical phenomena (for example waves). To explain physical phenomena more than one model may be needed.
4. $v = lf$, $d=vt$, and $f=1/T$ are mathematical equations that model the relationships between different wave characteristics.

What students should be able to do for any observable wave:
1. Obtain numerical values for l, v, f, T, and A by:
    - applying the definitions of l, v, f, T, and A and using appropriate measuring tools.
    - using the mathematical equations and known values to find the unknown values. (ex. Given l and f find v)
2. List, describe, and explain 3 different models of a wave.
3. Identify characteristics of a wave that are correctly and incorrectly represented by a particular scientific model.

One disadvantage of using guided inquiry lessons to present science content (instead of lecture presentations) is that students may miss the important concept they were to "discover" because they are so involved in the correct or incorrect "doing" of the activity. Even after a whole class debriefing discussion at the end of each lesson, much as we would like to believe our clear articulation of the subject matter is enough, examination of student notes in their workbook indicates some students will still be unclear. The "know and do boxes" make the learning objectives concrete and by stating them in terms of what students should know and be able to do, we clearly operationalize them for the student. The 12 "know and do boxes" written for the semester serve as another medium for getting the important points of the guided activities across and since the end of the unit exam questions are constructed from the "know and do boxes" students also find them useful as study guides.

Guided Inquiry Assessment Tools: The Unit and Final Exams

End of unit and final exams are used to assess the students' mastery of the learning objectives in the guided inquiry lessons. Some of the questions on the exam are fairly traditional for a conceptual physics course and consist of pencil and paper test questions requiring short answer explanations, drawings and calculations.

The Spring 2000 Exam 1 questions from unit one matched to the "know and do box" #1 illustrated above are shown below:

Questions from Physics 100 Exam 1 Spring 2000

1. Two children Jeremy and Jennifer are playing "high water low water" on the playground with a jump rope. The children are standing **3 meters** (300 cm) apart from each other. The jump rope has a knot in it, midway along its length. **Jeremy** shakes the rope up and down very hard to produce several waves. Here is a snapshot of what the rope looked like.

Jeremy   sJennifer

a) On the diagram above label one wavelength of the wave. Calculate one wavelength of the wave produced by Jeremy

b) In the picture, you can see the knot in the rope. The knot travels up and down (one complete cycle) in 2 seconds. Calculate the frequency of the wave.

c) Calculate the speed at which the waves move along the rope.

d) On the diagram above label the direction of energy travel and label the direction that the medium is vibrating.

e) Is this wave longitudinal or transverse?   Circle one: longitudinal or transverse
Explain your answer using a diagram.

---

5. Any of the following can be used to model a sound wave:
    a) a compression wave in a slinky (made by compressing the coils on one end)
    b) a water wave created by dripping water into a pan of water
    c) the wave equation $v = lf$
    d) computer program Kaboom
    e) the wave machine made of straws and fishing line
    f) wave in a rope

i) List a model that correctly models the longitudinal nature of a sound wave. Explain how it demonstrates this wave property.

ii) List a model that correctly models the waves that are present in a panpipe. Explain your choice.

iii) Select one of the models above and discuss one way in which the model does not accurately represent a sound wave.

Indicate model selected:
Discussion:

In addition to these pencil and paper questions there is at least one problem on the exam that requires students to manipulate materials and measuring devices used in the guided inquiry lessons. For example, in the waves and sound unit, students are asked to use the computer program "Kaboom" to measure the frequency of a sound wave produced by blowing across the top of an empty soda bottle. In the electricity unit students are asked to construct a circuit with a battery and light bulbs and make current and voltage measurements.

Open Inquiry Management Tools: The Interactive Journals

Picture this: Six groups of students are working on different inquiry questions, each with different group dynamics, uncertainties about experimental procedure, questions about key concepts and interpretation of data. Each group needs your immediate attention and you have 30 minutes to help them all. Impossible? Yes, it is! This is the situation we found ourselves in when we piloted the open inquiry component of the course. Luckily there were three instructors during the pilot course and each group was attended to, but this was a luxury we could not count on for future semesters. We had a classroom management problem that needed to be solved if the open inquiry component was to survive beyond the pilot phase.

The use of interactive group journals turned out to be the key to managing this communication problem. At the start of the inquiry when students are discussing and deciding on a question they are given a folder to journal the progress of their inquiry and ask questions. The inquiry journal contains:

- The individual and group inquiry planning sheets
- A group organization sheet designating group responsibilities and an attendance record
- Daily progress reports on the inquiry that include: responses to instructor queries, diagrams of experimental set ups, data, background literature research, questions for the instructor, analysis of the data and action plans for the next inquiry meeting.
- Questions from students, requests for materials or references

A few pages from Diana, Neng, Jenn and Andrea's inquiry journal are reproduced on the next page followed by a short description of its contents. The complete journal consisted of 29 pages of handwritten notes and diagrams and 29 pages of background research. Most of the background research was obtained and printed from the Internet.

This is the same group shown in part III of this paper. Their inquiry question was:

> **Does the pH level in different fruits affect their voltages?**

Pages from an interactive student inquiry journal

Page 1

Inquiry Group Organization

♣ **leader**   DANA
- responsible for keeping the group focused on the project and making decisions about the direction of the inquiry

♥ **recorder**   ANDREA
- responsible for keeping the inquiry journal which includes:
  - writing journal entries after each class period spent on the inquiry
  - documenting experimental set up, procedure, and data

♦ **materials manager**   NENO
- responsible for ensuring that all experimental equipment is available for the inquiry and is returned at the end of class.
- responsible for ensuring that adequate background literature is available for the inquiry.

♠ **presenter**   JENN
- responsible for assembling the poster presentation. All text must be typed and graphs and charts should be computer generated when possible.

Inquiry attendance Record

Please have each group member initial every day that they were in class to work on the inquiry.

Name	3/28/00	3/30/00	4/4/00	4/6/00	4/8/00
[illegible]		DG	DH		
DANA	✓	✓	✓	✓	NC
NENO	✓	NC	N.	NC	NC
ANDREA	✓	AC	Pr.	AC	AC
JENN	✓	SD	SD	SD	SD

Journal checklist:
- ☑ Name of Inquiry
- ☑ Names of group members and group role assignments
- ☑ Inquiry Planning Sheets: one for each individual and one for the group question
- ☑ Copies of background research and references used
- ☑ Data Table
- ☑ Raw data (ex. actual Kaboom recordings or multimeter measurements)
- ☑ Daily progress reports which will contain your "journal evidence".

Page 8

JOURNAL ENTRY #7

- Everyone brought the required materials.
- Pineapple was omitted because it was too expensive ($5).
- Crushes were obtained & softwire on all pH tape was obtained at the Depot or teacher's supply store. We found screws obtained @ Home Depot for each.
- apple, mango, grapefruit, tomato, kiwi, lemon, lime, banana

Everyone chipped in $1 to cover the cost of materials.

Questions:
1) Why do we need Cu + Zn? Why do metals work?
2) What does pH have to do w/ voltage? i.e. the lower the pH, the higher the voltage.

Both questions for the inquiry

Page 9

Procedure

Take each piece of fruit.
Place Cu screw in one end.
Zn screw in other.
Make sure screws are about 5 cm apart from each other + about 3 cm deep into fruit.

(R) wire — Cu   (B) wire — Zn
multimeter

Results

	Volts	pH
Grapefruit	.96 V	2 pH
Kiwi	.84 V	3 pH
Lime	1.00 V	1 pH
Lemon	1.01 V	1 pH
Orange	.95 V	3 pH
Tomato	.84 V	4 pH
Apple	.97 V	5 pH
Banana	.83 V	5 pH

Page 10

Plan

- Jenn is going to be the official typist.
- She is typing p materials, title, group members, question, + raw data.
- We are all going to do some research on pH levels and how it correlates to acidity level, + also why we needed to use Cu + Zn screws

NEXT MEETING

Formalize our analysis. Write up process, analysis, and conclusion.

Decide on poster format.

1) We plan to buy our own poster board + put together our poster at a later meeting. Everyone just likes coming from other sources that run good start - data looks good.

Questions:
2) Did you evaluate the source usefulness and accuracy of the background data you included? Did you just settle for looking up a holiday light? And why not a try searching under subject area & education, K-12 or university data. Try Val, Lemon Battery Science studying the net.
3) The lemon pen may hold the key to understanding your pH level is ___

Description of journal contents:

Page 1 of the journal is the organizational page. It is used to assign or record group responsibilities and keep an attendance record.

Page 8 contains a project status report and questions that the group has raised:

> **Questions from students:**
> 1. Why do we need Cu + Zn? Why do these metals work?
> 2. What does pH have to do w/voltage?
>   lower the pH, the higher the voltage

Page 9 contains a description of the experimental procedure and data taken on that day.

Page 10 contains plans for the next inquiry meeting and the instructor's questions and comments.

> **Questions and suggestions from the instructor:**
> 1. Did you evaluate the source, usefulness and accuracy of the background data (and research) you included?
> 2. Try searching under subject area of education, K–12 or university and key word lemon battery if you are surfing the net.
> 3. The Rxn may hold the key to understanding why pH level is important, consult a chemist if possible, Professor Londa Borer would be good.

Notes about the journal contents:

- Instructor question 1 was motivated by the background literature included with the journal. Portions of the literature were inaccurate and misleading and other parts were unrelated to the experiment.

- Instructor suggestions 2 and 3 were to help students locate more reliable and relevant resources for finding answers to the questions on page 8.

- An interesting observation: Most students used the computer as their primary source of background information. Some used CD ROM encyclopedias and most used the Internet to locate information. Almost none of them used the campus library or other available textbooks as resources.

The inquiry journals are collected at the end of each class period students are working on the inquiry and returned at the beginning of the next period. The time spent on giving students feedback is time well spent as they are eager and interested to read the comments made in the journal and are quick to respond to questions and suggestions. This is in contrast to the time I used to spend on grading weekly lab reports. My observation was that students would look at their grade and comments, but there was little indication that the grade or comments improved or encouraged student learning.

From the example above you can see that the journals are not formal laboratory notebooks. They are a place for students to think out loud and test their ideas. Their most important function is to keep the instructor informed of the direction of the inquiry and thinking of the group. With the use of the inquiry journals, students are able to work with minimal assistance during the class time designated for inquiry projects. There will occasionally be a group with a problem that is too big to be addressed by the journals, but these situations can be facilitated during the 30-minute inquiry period!

Open Inquiry Assessment Tool: The Open Inquiry Rubric

Assessment of the open inquiry projects is based on the journal, the poster presentation and peer evaluations. The journal and the poster are graded using a rubric. Being an active facilitator of the inquiry projects and observing the amount of effort and time students put into their inquiry projects can make it difficult to be an objective evaluator. Using a criterion-referenced rubric helps to objectify the grading and also gives students clear information as to what is expected of them. Students are given these rubrics at the start of the inquiry project so the rubrics also serve to manage student expectations of how the project will be graded.

To guide student and instructor thinking there is a rubric for each stage of the inquiry:

- The Question
- The inquiry process — The Inquiry Journal
- The poster

After our observation of several inquiry cycles we recognized the importance of developing collaborative work skills. Much has been written about the benefits of group work. However, group dynamics can be tricky to manage and students, even adult students, need some guidelines for working in a group. Since the open inquiry projects depend heavily on the ability of the group to function collaboratively, we felt the need to articulate some guidelines for behavior in a group. The last rubric is an attempt to describe some of the observable characteristics of a functional collaborative learning group.

P100 Inquiry Project Grading Criteria

The Question	Getting There	Got It!	Wow!!
Scope and Testability	• The question is too broad (it cannot be answered within the scope of this class) or too narrow (it can be answered by a quick test or by reading a reference book). • The question does not have a possible answer that can be verified by an experiment or test.	• The question relates to concepts in the course and requires some investigation or experimentation. • The question is testable. An experiment with well-defined and controlled variables can be designed.	• The question leads to a deep investigation of the course content and extends classroom learning by making connections to other science disciplines or phenomena outside the classroom. • The question can be tested in more than one way or on more than one level of accuracy or sophistication.

The Process — The Inquiry Journal

Background Research	• Uses only hearsay or personal opinion. • Uses only one reference or resource. (For example, just a textbook.) • Reference makes incorrect or superficial connection to the inquiry question. *Journal evidence:* • *No copies of background research materials are provided or research materials make no connection to the inquiry;*	• Uses at least two reliable resources among science reference books and reputable internet sites. • Key science concepts are identified and references make connections between these key science concepts and the inquiry question. *Journal evidence:* • *Copies of background research necessary to analyze the question are provided and important passages or facts are highlighted.*	• Uses more than two resources, including science reference books, science journal articles, reliable internet sources and interviews. • References are used to construct correct understanding of key concepts. This understanding is used to explain the observation or results of the inquiry experiment. *Journal evidence:* • *Copies of background research are provided and important passages or facts are highlighted. In their analysis, students restate these passages in their own words and use the information to analyze their data and answer their inquiry question.*
Design of Quantifiable Test or Experiment	• The test did not measure anything. Results were not quantifiable as a table or chart. • The test had conditions or variables that could not or were not controlled. *Journal evidence:* • *Data was gathered but it did not provide any information on the question.* • *Variables that might affect the outcome of the experiment were not identified or discussed.* • *The experimental design was poorly described or illustrated.*	• The test used tools to make at least one measurement. Data was gathered that could be displayed in a chart or graph. • The test identified variables that were to be held constant and variables that were to be changed. *Journal evidence:* • *Data was recorded that provided information on the question.* • *Variables that might affect the outcome of the experiment are identified and controlled systematically.* • *Experimental design was clearly described and illustrated.*	• The test had more than one layer to it, or looked at the problem from more than one angle. More than one tool was used to make measurements or more than one test was designed resulting in more than one set of data to be graphed or charted. • Tests demonstrated understanding of a multi-variable problem by careful control of experimental conditions and variables. *Journal evidence:* • *Data was recorded from several trials or several different tests.* • *Variables that might affect the outcome of the experiment are identified and controlled systematically.* • *Several experimental designs were tried. Each design was clearly described and data collected was analyzed and used to improve the design or to design an additional test.*

~ 29 ~

The Process — The Inquiry Journal continued

Experimental Procedure and Analysis	• A single test or experiment was conducted one time. The results were accepted, and a conclusion drawn based on one test. • Data was inaccurately or incompletely collected. *Journal evidence:* • Only one set of data is recorded. • There is no description of the conditions under which the data was collected. • There is no attempt to "make sense" of the data in the context of background material or classroom concepts.	• The experiment was conducted and analyzed more than once. • A dialogue about the question, experiment, results and background information led to the refinement of the question or experiment. • Enough data and background information were accurately collected to provide evidence to support an answer to the question. *Journal evidence:* • Data from several trials of the same experiment were collected. • A narrative of what students thought about their data and why they thought it (analysis) connects the data to background research and classroom concepts. • Conclusions and modifications of the experiment as a result of analysis are recorded.	• More than one test was conducted and analyzed more than once. • Much dialogue led to testing the question in a wide variety of ways, cycling often between testing and refinement of the question. • Data and background information was collected to support the answer to the question and led to a correct and thorough understanding of the key science concepts required to understand and explain the inquiry. *Journal evidence:* • Data from several trials or different types of experiments is recorded. • A narrative of what students thought about their data and why they thought it (analysis) connects the data to background research and classroom concepts. • Modifications of the experiment as a result of analysis are recorded and implemented. • A conclusion and answer to the inquiry question that is supported by the analysis is written.
Questions and Instructor's Comments	• Inquiry group did not respond to instructor's comments or questions on the inquiry. *Journal evidence:* • There were no written responses to instructor's request for clarification or explanation of journal entries.	• Inquiry group responded to instructor's questions or comments. *Journal evidence:* • There were written responses to instructor's request for clarification or explanation of journal entries.	• Inquiry group responded to instructor's question or comments and used the journal to pose questions and ask for help. *Journal evidence:* • There were written questions as well as responses to instructor's queries.

Journal Checklist:

- ❏ Name of Inquiry
- ❏ Names of group members and group role assignments
- ❏ Data Tables
- ❏ Inquiry Planning Sheets: one for each individual and one for the group question
- ❏ Copies of background research and references used
- ❏ Raw data (ex. actual Kaboom recordings or multimeter measurements)
- ❏ Daily progress reports that will contain your "journal evidence."

The Poster Presentation

	Getting There	Got It!	Wow!!
Organization and Appearance	• The presentation does not contain all key components in the checklist below. • The presentation is sloppy or illegible.	• The presentation contains checklist items, and they are all clearly labeled. • The presentation is neat in appearance. The text and most of the graphs and charts are typed or word processed.	• The presentation contains checklist items, and they are all clearly labeled. • The presentation is professional and interesting in appearance. All of the text and charts are computer generated and graphics or photos are used to enhance the presentation.
Procedure	• There is no description or an incomplete description of how and why the data were collected. • From the description it is not possible to understand the relevance of data to the question.	• There is a description of the experimental procedure and a list of the materials used. • From this description it is possible to see how the data is related to the inquiry question and how it might answer it.	• The description of the experimental procedure was a step-by-step account of what was done and why and what materials were used and where they came from. • From this description the experiment and its data could be reproduced by someone else.
Graphs and Charts	• Graphs or charts are missing or are incorrectly or incompletely labeled. • Graphs or charts do not represent data that is relevant to the inquiry question.	• At least one graph and one chart are presented and correctly labeled. • At least one graph or chart represents data that provides insight into the answer to the inquiry.	• Multiple graphs and charts are presented and all are correctly labeled. • All graphs or charts presented provide visual evidence of the answer to the inquiry.
Analysis (What you thought of your data and why)	• Analysis is missing or incorrect. • Explanation or discussion of data show little evidence of logical reasoning. • Explanation of data doesn't correctly identify or use key concepts.	• Explanation of the data is made with logical reasoning and based on the correct understanding of key science concepts. • Explanation connects background research to data collected.	• Explanation of the data is made with logical reasoning and based on the correct understanding of key science concepts. • Explanation also addresses the limitations of the experiment and estimates the accuracy of the data collected (i.e., is able to give a + or - estimate of error on measurements made).
Conclusion	• The conclusion simply restates the data or analysis or says that they were "right" or "wrong" in their thinking.	• The conclusion answers the question and makes a statement about the significance or importance of the answer.	• The conclusion answers the question and makes a statement about how the answer generalizes to another phenomena or relates to an event outside of the classroom.

Presentation Checklist:

- ❏ Name of Inquiry
- ❏ Names of Group Members
- ❏ Inquiry Question

- ❏ Graphs and Tables of Data
- ❏ Description of Procedure
- ❏ Data Analysis

- ❏ List of References Used
- ❏ Conclusion

The Process — Group Work — The Inquiry Journal

Management of the Cooperative Group	• Group members did not take responsibility for their assigned roles. Some group members were absent or otherwise failed to contribute to the group • The level of involvement and work load were not evenly distributed. One or two people in the group did all of the decision-making and work on the inquiry. *Journal evidence:* *• There is no evidence of group planning. No written plan for what will be done next to make progress on the inquiry.* *• Group planning is vague and does not specify who will do what for the next inquiry meeting.*	• Each group member fulfilled his or her assigned role in the group. All group members contributed to the design of the experiment, collection, analysis and presentation of the inquiry data. *Journal evidence:* *• There is some evidence of group planning. There are written plans for what will be done next, but it is not clear who will be responsible for elements of the plan such as background research, material gathering (if you need something special not furnished in the classroom), etc.*	• Each group member took responsibility for his or her role in the group and involved the other group members in decision-making responsibilities and work to be accomplished associated with that role. (For ex. The presenter solicits input on what the poster should look like and assigns each group member to construct a piece of the poster presentation rather than constructing the entire poster alone. • The workload was evenly distributed within the group. *Journal evidence:* *• There are written plans of what should be done next and who will do them. This list includes items such as background research, experimental design, collection of data, etc.* *• Each group member has a responsibility for the next inquiry meeting.*

The open inquiry rubrics have been a very useful tool for defining what we want students to experience and learn in the open inquiry process. The construction of these rubrics has forced us to describe what type of student work constitutes evidence of this experience and learning. Although the rubric will continue to be edited and refined, the many discussions among the collaborators have produced a rubric that adequately describes the evidence of student learning that we value in the open inquiry projects. We also have noted that as we more clearly articulated our expectations and what we value in the open inquiry projects to our students, the quality of the inquiry projects has steadily increased and more closely aligned with instructor expectations!

V. Evidence of Student Learning: What are Students Learning in the Course?

This is a big question, and it is one that can be answered by an infinite number of investigations. We have just started the process of evaluating the student experience and learning in Physics 100. Below are listed some evaluation questions we have begun to investigate and the data we have collected.

1. **To what extent have the inquiry based activities we designed enabled students to develop the ability to pose testable questions and design experiments to answer them?**
- We have collected students written questions and investigation plans at five points during one semester to chart the development of this ability (54 students).

2. **At what level of sophistication have cooperative groups generated meaningful conversations about science?**
- Audio tape recordings of open inquiry group discussions and student interviews have been collected for discourse analysis. Data collected over one semester for two groups (nine students).

3. **What are the students' perception of and attitudes towards the open inquiry experience?**
- Pre- and post-attitude surveys have been collected over three semesters for 240 students.
- Student self assessment data after the first inquiry project has been collected over two semesters (120 students).

The following discussion focuses on the first part of evaluation question 1:

To what extent have the inquiry-based activities in Physics 100 enabled students to develop the ability to pose testable scientific questions?

Introduction: What is the motivation for looking at students' ability to write testable questions?

Central to scientific inquiry is the concept of a question. The most useful questions in experimental science are testable questions also known in education literature as "operational questions"[9-11]. The ability to ask a testable question is the most crucial part of scientific inquiry[12] and a prerequisite to engaging in inquiry based learning. Being able to write a testable scientific question is a learning objective of the open inquiry instruction in Physics 100. To understand how experience with the open inquiry process and direct instruction affect students' ability to write a testable question, we conducted the following investigation.

Method:

To document students' ability to write testable questions throughout the semester we collected questions from each student at five points in the semester. These points were located before and after open inquiry projects and before and after direct instruction. There were two sections of Physics 100 and we collected five sets of questions from each section for a total of 247 questions. We then coded each student question as testable or not testable and examined the percentage of testable questions that were asked in each question set.

Procedure for collecting data:

Written questions and investigation plans from two Physics 100 sections (n=53 students) were gathered at five points during instruction in the spring 2000 semester. Questions were gathered using two methods. Three sets of questions were gathered using the Inquiry Project Planning Sheets and two sets of questions were collected using end-of-the-unit and final-exam questions. The following table compares and contrasts the two collection methods:

Method 1 Collection of student questions using Inquiry Project Planning sheets	Method 2 Collection of student questions using end-of-the-unit and final-exam questions
• Question was constrained to the general topic area being studied (waves and sound or electricity and magnetism or light and color)	• Question was constrained to phenomena selected by instructor. (electromagnets or the interaction of light with plants)
• Students were asked to write their question after two weeks of guided inquiry lessons on the general topic	• Students were asked to write their question at the conclusion, four weeks, of the guided and open inquiry lessons on the general topic.
• Students were given four days to write their question.	• Students wrote their question as part of a one hour and 40 minute or two hour exam.
• Students wrote their question in the context of an investigation plan.	• Students wrote their question in the context of an investigation plan.

Method 1, which uses the inquiry project planing sheets to collect student questions, is only used once in each topic area. Using this method more than once in a topic area would make it difficult to distinguish between an original student question and a question borrowed from a completed inquiry project since students will have observed multiple questions and completed inquiry projects in the topic area at the conclusion of that unit. For this reason Method 2 was used to collect additional questions and investigation plans from students. By constraining the questions to phenomena or topics not covered by completed student inquiries, in this case electromagnets and the interaction between light and plants, we could ensure the questions were conceived by the student and not borrowed from a previous inquiry.

A sample of the written materials used in Method 1 and Method 2 are shown below:

Method 1	Method 2

Inquiry Project Planning Sheet
 	
Name(s):
Inquiry Question:
 	
What will you measure and what will you measure it with?
 	
What will you graph?
 	
List of things you will provide for the inquiry:
 	
List of classroom materials you would like to use. (You may use any materials you have seen in class provided they are returned at the end of the class period. If you need something special consult your instructor):
	Final Exam Question Spring 2000
	
	6. Light interacts with living things. In particular green plants need light to live and grow. (The process of turning light energy into carbohydrates that plants need to live and grow is called photosynthesis.) Materials are provided to help you think of a testable question about the interaction between light and plants.
	
	Materials provided* Additional materials:
	    Lamp
	    2 light bulbs of different watts
	    colored filters
	    ruler
	    green plant
	    spectroscope
	*Note: you are not limited to the above list of materials. If your question or investigation requires additional materials please list and describe them above.
	
	a) Examine the materials and write one testable question about the interaction between light and green plants.
	
	b) Briefly describe your investigation plan.
	
	List variable(s) you will hold constant:
	List variable(s) you will change:
	List variable(s) you will measure:
	
	c) What will you measure and what will you measure it with?
	
	d) Illustrate a graph you might use to present your data. <u>Clearly label your graph axis.</u>

Experimental Treatments (Instruction) Affecting Students' Questions

We will examine the effect of open inquiry and direct instruction on students' ability to ask testable questions. The open inquiry process has been described in section III of this paper and the direct instruction on testable questions is described below:

Direct instruction on testable questions:

Although direct instruction on testable questions occurred over three class periods, the total class time spent on direct instruction was only about 20 minutes. Just before the third inquiry project planning sheet was to be completed, students were given 26 questions for homework (compiled from previous P100 inquiry planning worksheets) on electricity and magnetism and instructed to classify them as testable or not testable. They were also asked to identify common characteristics among the testable and non-testable questions. On the next day of class 20 minutes were spent discussing characteristics of testable and not testable questions. On the third day, a summary of that discussion in the form of the revised open inquiry question rubric shown below was handed out to students.

Getting There	Got It!	Wow!!
• The question is too general. It is not focused on a single concept, phenomena or observation. or • The question is too narrow. It can be answered by a quick test or by reading a reference book. or • The question does not have a possible answer that can be verified by an experiment or test. Most questions that begin with "why" cannot be answered with a test.	• The question is focused on a particular concept, phenomena or observation. • The question requires some investigation or experimentation. • An experiment with defined and controlled variables can be designed.	• The question requires investigation or experimentation where data can be collected. • Specific experimental variables are identified in the question.

Method of Determining the Testability of Students' Questions

The criteria and procedure for determining the testable or not testable classification of a student question was developed by Lynn, Patty, David Jelinek (CSUS Associate Professor of Science Education and NSF external project evaluator) and Professor Ron Tanaka (CSUS Professor of English).

Although questions had been collected in the context of planning inquiry investigation, the questions were coded out of context without reference to the proposed investigation plan. Lynn, Patty and David coded 73 student questions as testable and not testable according to the following criteria:

Testable and Not Testable Question Criteria

Testable: A question is testable if an experiment might be set up to answer the question. Not Testable: A question is not testable if an experiment cannot be set up to answer the question.

Testable or Not Testable classification was *not* based on the following:
1. How narrow or broad the question was
2. How easy or difficult the implied experiment might be
3. How important or interesting the question was

Although questions were reported as either testable or not testable for this investigation the questions were also coded into the following subcategories for future analysis.

Testable Question Subcategories:
T as is = Question is testable as it is written. It explicitly identifies independent, dependent variables and criteria for measuring or evaluating the outcome of the implied experiment. T reph = Question is testable. It may need to be rephrased for clarity or to explicitly state variables or measurement criteria. T too narrow = question is testable and will add to a students' experiential base of knowledge but is too narrow or shallow to lead to or improve conceptual understanding. A quick experiment or observation will answer this question. It is too narrow to lead to an inquiry project. T no mat = Question is testable, but materials or equipment is not available or practical in the context of a classroom. T concept error = Question is testable but it assumes or is based on an incorrect conceptual understanding. This misconception can be clarified by an experiment.
Not Testable Question Subcategories:
NT why = Question is not testable because it asks "why" something happens or exists.
NT what is = question is not testable since some "what" questions cannot be answered by a test (i.e, what is an echo?). Asking what something is is asking for a description of a phenomena or an observation.

NT too general = Question is not testable because it is too general, it is not focused on a single phenomena or concept. It is unclear as to what the independent or dependent variables might be.
NT how = Question is not testable because it asks how something happens. Investigations are not usually done to answer "how" questions.
NT (mat) Question is not testable because materials do not exist anywhere to test the question.

NT concept error = question is not testable because it assumes or is based on an incorrect conceptual understanding that cannot be clarified by an experiment.
NT not question = what the student has written is not a question. For example a statement written with the intention of proving the statement with an experiment.

Of the 73 questions coded by Lynn, Patty and David there was agreement on the testability of all but one question suggesting a high interrater reliability among them. Fifty of the 73 questions were also coded by English Professor Ron Tanaka and his English 20 Critical Thinking Class. Interrater reliability was low between the English 20 group and Lynn, Patty, and David. This indicates that the criteria for testable and not testable questions is not definitively articulated. In applying the testability criteria the biggest problems we are aware of are:

- use of the criteria depends on the reader's ability or inability to envision the experimental setup required to answer the question
- the criteria do not address how much a reader might infer from a question that is unclear grammatically.

The only syntax pattern all the raters identified as important was beginning a question with the word "why." All raters concluded that questions that ask "why something is so" or "why something happens" cannot be answered by a test.

The Timeframe of Data Collection:
To understand the time frame of data collection and its relationship to instructional interventions, time is measured in units of class meeting days. Physics 100 had 31 class meeting days in the spring semester. (The course met twice a week for 15 weeks plus one day for the final exam.) The Inquiry Project Planning Sheet was used to collect questions on class days 5, 17 and 25. Written exams were used to collect questions on class days 22 and 31. Open inquiry projects were completed and presented on days 9, 21, 29 and 30. Direct instruction of how to write a testable question occurred over days 22, 23 and 24.

Below is a tabulated and graphed summary of the data collected:

Question Set	Method of Collection	Class Day #	Student experience at time of writing question	P100 Sec. #	Total # of questions collected	% of questions testable
1	1	5	No open inquiry experience	4	28	28
				5	24	24
2	1	17	1 open inquiry completed	4	24	24
				5	20	20
3	2	22	2 open inquiries completed	4	27	27
				5	25	25
4	1	25	2 open inquiries completed and direct instruction	4	23	23
				5	23	23
5	2	31	3 open inquiries completed and direct instruction	4	*	*
				5	*	*

*Data has just been collected and has not been analyzed

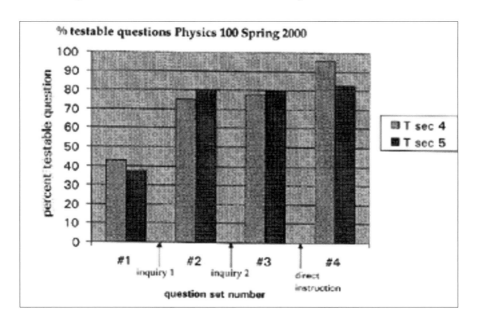

Preliminary Analysis:

Although the students had participated in four class periods of guided inquiry activities and were specifically instructed to write a testable question and plan an investigation to answer their question, less than 50% of the questions written were testable. After the first experience with open inquiry where students participated in the design, analysis, presentation and evaluation of a scientific investigation to answer a testable question, 75% of the questions written were testable. After a second experience with open inquiry there was a small increase in the percentage of testable question written, sec. 4 was 78% and sec 5 was 80%. We question whether this increase is statistically significant. After direct instruction on the characteristics of a testable question 96% of the questions asked by section 4 were testable (an increase of 18%). However only 83% of the questions asked by section 5 were testable (an increase of only 3%).

Tentative Conclusions

It appears that participating in guided inquiry lessons as we have defined them in Physics 100 does not enable most students to write a testable scientific question. Participation in group open inquiry project appears to improve some students' ability to write a testable question. Repeated participation in group inquiry projects without direct instruction does not appear to benefit students. Direct instruction as described in this study appears to improve the ability of some students, but may have little effect on others. Final data including question set 5 may provide more insight on our investigation.

References

1. McDermott, L.C., *Physics by Inquiry volumes 1, 2, and 3* (Physics Education Group, Physics Dept., University of Washington), New York: John Wiley & Sons, Inc., 1996.
2. McDermott, L.C., and P.S. Shaffer, "Research as a guide for curriculum development: An example from introductory electricity, Part I: Investigation of student understanding," *American Journal of Physics* **60** (11), 994 (1992); Erratum to Part I, *American Journal of Physics* 61 (1), 81 (1993).
3. Shaffer, P.S., and L.C. McDermott, "Research as a guide for curriculum development: An example from introductory electricity, Part II: Design of instructional strategies." *American Journal of Physics* **60** (11), 1003 (1992).
4. Hestenes, D., "Toward a modeling theory of physics instruction," *American Journal of Physics* **55**, 440 (1987).
5. Wells, M., and D. Hestenes, "A modeling method for high school physics instruction," *American Journal of Physics* **63** (7), 606-619, July 1995.
6. Karplus, R., "Science Curriculum Improvement Study: Teachers Handbook," Berkeley, CA; Lawrence Hall of Science, 1974.
7. Barba, R.H., *Science in the Multicultural Classroom A Guide to Teaching and Learning,* Chapter 8, pp 186 - 193, University of New Mexico, Needham Heights, MA; Allyn and Bacon, 1995.
8. *National Science Foundation, Division of Elementary Secondary, and Informal Education, Volume 2 FOUNDATIONS, Inquiry: Thoughts, Views, and Strategies for the K-5 classroom*, NSF 99-148, Chapters 5, 6 and 10, L. Rankin, B. Kluger-Bell, D. Ash.
9. Alfke, D., "Asking operational questions," *Science and Children*, **11**, 18-19, 1974.
10. Allison, A.W., and R.L. Shrigley, "Teaching children to ask operational questions in science," *Science Education*, **70**, 73-80, 1986.
11. Goh, Ngoh-Khang and Lian-Sai Chia, *Pre-Service Teachers' Behavior and Training in Asking Operational Questions*, Research/Technical report, 1992.
12. Roth W.M., and A. Roychoudhury, "The development of science process skills in authentic contexts," *Journal of Research in Science Teaching*, **30**, 127-152, 1993.

Curricula Gridlock and Other Challenges in Teacher Preparation

Karen L. Johnston
North Carolina State University

Viewpoint

John Dewey had it right. Education is the cornerstone of democracy. The responsibility for the education of young people resides with all of us. With purposeful attention and deliberate actions, we will match and even surpass the great strides we have made and the successes we have enjoyed in science and education during the past 50 years.

When Congress passed the National Science Foundation Act and President Truman signed this into law on May 10, 1950, the science (and science education) enterprise in this country became linked to the public in a real, measurable and accountable manner through public tax dollars. Initiatives on behalf of the science education enterprise emerged from the public voice for science. That public voice spoke of a need to "provide a nation with an aristocracy of talent."[1] Bright students needed to be encouraged to attend college; and science, physics in particular, needed to find ways to identify those best suited for advanced studies. Almost 50 years later, that public voice, recognizing the successes of a strong, albeit very different, science community, suggests a more populist perspective for science education. Shaping the Future[2] recommends a new vision, one proposing that "all students have access to supportive, excellent undergraduate education in science, mathematics, engineering and technology, and that all students learn these subjects by direct experience with the methods and processes of inquiry." Shaping the Future recognizes that we have communities of researchers and practitioners committed to the science education enterprise who are capable of and committed to making science available to all students. Our task is no longer limited to identifying and educating talented students as the centerpiece of science education, though we must continue to do so, but offering **all** students an education in science.

The student remains our focus during this transition period, and the teacher is the most influential connection to the student. Therefore, our attention to the preparation of K–12 teachers is more necessary than it has been in the past 50 years. Physics departments cannot simply discharge their duty by providing an occasional workshop or relegating a lone faculty member committed to K–12 education matters to offer a few courses for teachers here and there. Our standards of what is acceptable educational practice, shaped by the merger of research in developmental psychology and cognitive science with students' conceptions of physics, have changed. We recognize that new curricular materials developed by physicists and based on research are more effective in helping students learn physics. In addition, we support a cadre of physicists, from university and precollege settings, who are able to translate our new knowledge into practice. Considering all of this, we have what we need to do a better job of preparing future teachers.

It is the right time for every physics department in two-year colleges and four-year colleges/universities to take deliberate actions to develop coherent programs that prepare K–12 teachers who understand science and are skilled at making science a part of every person's life. So, let's get to work.

Student Learning and Standards: The Touchstones for Change

Almost 30 years ago *Physics Today* boldly published an issue that highlighted children learning physics. With articles by Jean Piaget[3] on developmental patterns of reasoning and Robert Karplus[4] on children's conceptions of science principals, the early efforts of physicists doing physics education research were validated. Karplus suggested we had a lot to learn from our students and could do so by probing their view of the physical world. He invited physicists to engage in this all-important task of linking science and the practices of science to the educational enterprise more directly and to expand the audience of students having access to science education. By keeping science education visible through curriculum development projects, teacher institutes and projects that capitalized on research into learning and later, new technology tools, we have positioned the physics community to respond to the ever-changing landscape of education.

The future seems promising. AIP[5] reports increases in student enrollment in high school physics. New curriculum projects designed to offer physics to ninth graders signal that the 50 years of post-Sputnik science education initiatives have produced some successes, and we are compelled to expand our attention to students beyond the small subset of those who will become practicing physicists. Physics departments have established research groups engaged in physics education, and while the number of these groups is small, it is growing. The scholarly work created by physicist-educators is expanding. One such example provides a perspective of physics education research (PER) by blending carefully selected constructivist language within a scholarly framework. Redish[6] captures what is at the heart of the National Science Education Standards (NSES or Standards)[7] in a few principles and corollaries in language a physicist can appreciate.

The NSES document is influencing the way we think about science education through the inclusive language in which it is written. The Standards demand we put less emphasis on the tired arguments engaged in by academicians from two different "camps" regarding whether the "s" or the "e" should be the focus of science education for teachers. Instead, we are compelled to focus on students' learning and teachers' skills at affecting that learning. The Standards declare that science teaching is about understanding science. The Standards demand that professional development programs engage teachers throughout their career. From the Standards we can infer how to measure and evaluate the quality of our science programs for teachers and students.

The NSES are based on five assumptions that form the foundation of **science for all**. These assumptions are listed in Table 1.

Table 1. Five assumptions on which the NSES for science teaching are based.[8]

Assumption 1:
The vision of science education described by the Standards requires changes throughout the entire system.

Assumption 2:
What students learn is greatly influenced by how they are taught.

Assumption 3:
The actions of teachers are deeply influenced by their perceptions of science as an enterprise and as a subject to be taught and learned.

Assumption 4:
Student understanding is actively constructed through individual and social processes.

Assumption 5:
Actions of teachers are deeply influenced by their understanding of and relationships with students.

The NSES model of professional development, buttressed by these five assumptions, occurs along a continuum from undergraduate studies throughout the professional career. [Studies undertaken by the National Institute for Science Education likewise support the concept of continuous professional learning experiences for teachers.[9]] In essence, the Standards instruct us that those who teach are *always becoming a teacher*.

The Standards and results from PER should not be unreflectively accepted, but should be deeply understood. They should change the way we view teacher preparation and the way we do our personal teaching. The Standards, coupled with the results from PER, serve as the two external guideposts for the development or improvement of professional development programs for teachers.

It is easy, however, to get caught up in discussions about the crisis related to the numbers of science teachers in the United States. It is these numbers and the projections of shortfalls that seem to provide the impetus for action. We fret about the pipeline. Are we preparing enough science teachers (output)? Do our programs prepare teachers in "the right way" (quality control)? How can we improve retention of teachers in the early career years (preventing leakage)? Are programs for crossover teachers doing the job (managing seepage)?

The pipeline is not the problem; the pipeline simply is. By turning our attention away from the numbers and toward the task of building and maintaining the infrastructure necessary to support the nurturing and developing of science teachers, we address the pipeline problem in a way that fits our professional role. University and departmental infrastructure provides content, coherence and stability for professional-development programs. When science faculty and administrators, education faculty and administrators, and K–12 teachers and school personnel work together, the efforts address the more important concerns of the teacher in the classroom and produce the desired outcomes, such as (1) developing meaningful coursework for pre- and in-service teachers, (2) continuing to do research in learning and disseminating the new knowledge, (3) developing scholarly collaborations between science and education faculty and (4) strengthening professional communities by linking all elements of the K–G science education enterprise.

When N is Small. . . the NC State Story

Two elements of the culture of North Carolina State University are the drivers for science-teacher professional development efforts in the Department of Physics: (1) the land-grant character of the university and associated mission of extension services (outreach) and (2) the absence of an elementary-education degree program. As a land grant Research I institution, the university is obliged to reach beyond the confines of the campus laboratories and classrooms to connect North Carolina citizens to the university and her work. In the last two decades, this extension mission has been embraced by the sciences, at first with individual faculty members undertaking outreach efforts on an ad-hoc basis and more recently through an interdisciplinary outreach center for the College of Physical and Mathematical Sciences.

The absence of an elementary-education program in the undergraduate curriculum results in a small number of preservice teachers enrolling in physics courses. The steady state number of those who plan to be science, mathematics, or technology (SMT) teachers in middle schools or secondary science education enrolled in introductory physics/astronomy courses in any academic year is less than 50. These two characteristics of the NC State culture are reflected in how we expend our efforts in K–12 teacher professional development. More effort and resources are directed toward in-service education, and the in-service efforts are consolidated with those of other departments in the College of Physical and

Mathematical Sciences (PAMS) in a resource facility called The Science House[10] located on the Centennial Campus of the university.

Professional Development Programs: Inservice

By consolidating the outreach efforts of the mathematics and several science departments, David G. Haase, physicist and Director of The Science House, guides outreach efforts in science and mathematics from an interdisciplinary perspective. Programs for middle and high school teachers and their students are offered year-round and made possible by having a permanent facility and a permanent professional staff of faculty and teaching specialists. Reaching over 600 teachers and 20,000 students annually, The Science House is, in every sense, a partnership between the university and K–12 teachers. Permanent in-service facilities and professional staff seem to be the infrastructure needed to establish and maintain the scholarly collaborations between the science disciplines, the education faculty and the partner schools. Table 2 illuminates the range of activities at The Science House. A more detailed description of each program is found at The Science House Web site.[10]

Table 2. The Science House—A Learning Outreach Program.

Resource	Description	Programs
Teacher Toolbox	Workshops in SMT, equipment lending library-CBL and MBL	EMPOWER
Stuff for Students	Science research experiences, SMT experiences for student groups under-represented in SMT	IMHOTEP Academy, Expanding Your Horizons PAMS Summer Camps
Curriculum Closet	Hands-on activity books in chemistry/physics, physics curriculum activities aligned with NC standards	Physics on the RoadFun with Physics (vt)
Science Junction	Cyber-community for teachers, students and researchers with collaborative experiments	Links to: SERVIT, NC State research, SciTeach Forum, Lesson Plans linked to Standards
Hands on Happenings	Up-to-date calendar of local and national SMT activities	

The workshop-related programs for teachers provide not only hands-on experiences with equipment and technology tools, but also encourage and assist teachers in building resources of science-teaching equipment at their own schools.[11] Existing and new initiatives are aligned with the NSES and state standards.

Preservice Teacher Education: Curricular Gridlock and Licensure Dilemmas

The teacher preparation model at most institutions in North Carolina is a clinical model[12] where faculty members who teach in the schools (K–12) are involved in the professional development programs for aspiring teachers. This clinical faculty offers their expertise in methods courses, supervision of practicing teachers and mentoring of new teachers. At NC State University, veteran teachers serve as mentors for new teachers in the induction years. Other clinical experiences include teaching practice opportunities followed by guided reflection on the experiences and skill development that includes demonstration and practice followed by feedback. The NC State program is accredited by the National Council for Accreditation of Teachers (NCATE).

Preservice teachers in SMT are enrolled in degrees programs that fulfill state or NCATE requirements for licensure either at the middle-school level or secondary level. The Science Education (SED) program for grades 9–12 prepares prospective teachers for licensure in **comprehensive science**, meaning that the teacher has the credentials to teach any science taught at the high-school level. Licensure in comprehensive science requires prospective teachers to select a science concentration and to take at least two courses in each of the other three areas of science. For example, a physics concentration requires approximately 30 semester hours of physics and two courses each in chemistry, earth sciences and biology. None of the physics courses in which an SED student would enroll are specialized courses for preservice teachers, and many of the lower division courses would be properly characterized as traditional large lecture section courses. The upper division courses for SED students in a physics concentration are the courses regularly taken by physics majors. With never more than one or two SED students per semester and rarely a student who has selected a physics concentration, specialized courses in physics for SED students are not fiscally feasible.

Until recently, the undergraduate physics program at NC State was a rigorous Bachelor of Science (BS) degree program providing opportunities for research experiences. During the past five years, physics department adopted and had approved a more flexible Bachelor of Arts (BA) curriculum to complement the existing BS degree. The BA degree provides ways for students to express interest in other technical areas without sacrificing the rigor of the undergraduate physics program. Although we had no expressed demand for preparing high school physics teachers with a major in physics, we took this occasion to work with faculty from the Department of Mathematics, Science and Technology education to explore the possibilities. This exercise revealed how crowded teacher-preparation programs have become with requirements. For example, future teachers now must take courses in multicultural education, tutoring adolescents and teaching exceptional students. In addition, a workable program had to include two courses in each of the three other science areas since licensure was for comprehensive science. The physics departmental course and curriculum committee debated the merits of considering science education courses as **technical** electives and in the end were convinced that these courses, indeed, were the technical background of the teacher-preparation program. The final product is a dual degree (BA physics/BS science education) requiring four years plus one summer. While both departments streamlined the program, it offers little breathing room for the student.

NCATE recently issued a set of more rigorous performance-based standards requiring teacher candidates to have a major or the "substantial equivalent" of a major in their area of expertise.[13] Whether these new standards will increase interest and demand for our dual degree program in physics and science education remains to be seen.

The number of students enrolled in physics courses seeking **middle-school licensure in science**, averages around 10 students per semester. A student seeking middle-school licensure with a science concentration would take two semesters of physics, such as a one-semester conceptual physics course and one semester of introductory astronomy (in addition to science courses other than physics). Middle-school licensure with a concentration in both mathematics and science requires two semesters of college-level physics (in addition to other science courses). All of the degree/licensure programs require a minimum of 128 semester hours, with 16 of those hours reserved for the methods/student teaching experience. Do middle school science teachers need more physics or more specialized physics courses? Most likely. Unfortunately, to meet the demands of comprehensive science licensure, university-required general-education courses, education courses demanded by NCATE and the sampling of courses from four science areas, the curriculum has no room for more physics. The small number of SMT education students we encounter in our courses makes it unfeasible to offer for teachers the specialized courses that are demonstrably more effective than traditional introductory physics.[14]

Our most promising solution to the small N and gridlocked curricula problems for students in preservice education programs relies on insuring that introductory physics and astronomy courses are more closely aligned with instructional methodology promoted by NSES. The PER and development group in our department are engaged in initiatives to do just that. Under the leadership of Robert Beichner, the university is building a state-of-the-art interactive classroom that will integrate lecture and laboratory and accommodate the large sections of introductory physics. The classroom is designed for collaborative work by students, making it easy to move from low technology, hands-on experiences to more high-technology tools for addressing real-world problems. Recognizing that teachers influence the achievements of their students far more than any other observable variable, discussions have explored ways in which SMT education students (both middle and secondary) can be directed to the specific physics classes that are conducted in the interactive classroom. In addition, the department is taking the lead at the university to assemble an interdisciplinary group of SMT faculty to focus on teacher professional development.[15] The premise for this collaboration is that research will drive permanent changes in the courses we teach and improve the ways in which we prepare SMT teachers. External funding provides opportunities for change and can accelerate change, but it is faculty, students and university infrastructure that implement and make permanent those changes.

Advice: Begin with the End in Sight

The essential elements for developing effective teacher professional development programs are to: (1) involve all stakeholders in that effort, (2) buttress the program with the scholarship of learning and teaching and (3) build the institutional infrastructure that supports continual, incremental change.

Our collective wisdom and results of research tell us clearly that lecturing about physics does not develop understanding for most students. Lecturing about the importance of and how to do inquiry does not help teachers develop the understanding and skill to teach by inquiry. Walt Whitman says it well in "When I Heard the Learn'd Astronomer"[16]

> *When I heard the learn'd astronomer*
> *When the proofs, the figures were ranged in columns before me,*
> *When I was shown the charts and diagrams, to add, divide, and measure them*
> *When I sitting hear the astronomer where he lectured with much applause in the lecture room*
> *How soon unaccountable I became tired and sick,*
> *Till rising and gliding out I wander'd off by myself,*
> *In the mystical moist night-air and from time to time,*
> *Look'd up in perfect silence at the stars.*

If we want students to understand science one by one, then we must teach science in that way.

What are the ways and means to mend the fractured environments between K–12 institutions, institutions of higher education and professional communities? For program coherence, a professional development program at a university must have integrated preservice and in-service components that have been informed by research and the Standards. Faculty in both the education and science disciplines must work together respecting the expertise that each brings to the table. Administrators must be leaders of reform in the teacher professional development by supporting the requirements of these new faculty initiatives. Programs must have permanent space and other related resources. Faculty who devote their talents to teacher preparation and the related scholarship must be rewarded for their efforts. Teaching specialists from the K–12 classrooms must be invited into the university, on equal footing, to work as scholars with university faculty. Professional associations must develop on-going programs that engage pre- and in- service teachers in the larger professional communities. While the historical traditions and culture at the university must be respected, they need not interfere with or prohibit physics departments from actively pursuing a path that leads to science for all students.

Beginning with the end in sight means that each of the stakeholders must establish goals and identify benchmarks for reaching the goals. Continual program monitoring and evaluation protocols need to be established so that we make every attempt to measure our progress and gauge whether our actions are producing the desired effects. For example, can we demonstrate that our courses do improve student understanding of key physical concepts? Can we document that our pre- and in-service teachers are skilled in using the equipment and tools of technology as a result of our courses or workshops? Do our in-service efforts have an effect on teacher retention? Does active engagement in professional associations have an effect on teacher retention? Can we document that we teach our university physics classes by more interactive methods and what effect does this have on student learning and student attitude?

Holding our institutions, our departments and ourselves to high standards is what guides change and allows us to begin with the end in sight—science for all students. John Dewey had it right.

REFERENCES

1. England, J. Merton, *A Patron for Pure Science—The National Science Foundation's Formative Years, 1945–57,* National Science Foundation, Washington, DC, 1983, p. 228.
2. *Shaping the Future*, National Science Foundation, 1996.
3. Piaget, Jean, *Physics Today*, June 1972.
4. Karplus, Robert, "Physics for beginners," *Physics Today*, June 1972.
5. Neuschatz, Michael, et al., "Maintaining the momentum—high school physics for a new millennium," *AIP Report* Publication R-427, August 1999.
6. Redish, Edward J., "Implications of cognitive studies for teaching physics," *American Journal of Physics*, Vol. 62 (9), September 1994.
7. National Science Education Standards, National Research Council, 1996.
8. National Science Education Standards, National Research Council, 1996, p. XX.
9. Mundry, Susan, et al., "Working toward a continuum of professional learning experiences for teachers of science and mathematics," *Research Monograph* No. 17, National Institute for Science Education, 1999.
10. http://www.ncsu.edu/science_house.

11. Team Science, David G. Haase, et al., Final Report of a National Science Foundation Teacher Enhancement Project, 1997.
12. Learning to Teach in North Carolina, The University of North Carolina General Administration, 1992.
13. http://www.ncate.org/2000/pressrelease.htm.
14. McDermott, Lillian C., "A perspective on teacher preparation in physics and other sciences: the need for specialized science courses for teachers," *American Journal of Physics*, Vol. 58 (8), August 1990.
15. Beichner, Robert J., "Specialists and Teachers Advancing Reform," internal document.
16. Foerster, Norman (ed.), American Poetry and Prose, 5th edition, Houghton Mifflin, Boston, 1970, pp. 773-4.

How Computer Technology Can be Incorporated into a Physics Course for Prospective Elementary Teachers

Fred Goldberg
San Diego State University

Introduction

At San Diego State University we have developed a one-semester physics course for prospective elementary teachers that uses the same pedagogy and computer-based materials as developed for the NSF-supported Constructing Physics Understanding in a Computer-Supported Learning Environment project (or CPU Project).[1] The course is structured to support a learning environment where students take primary responsibility for developing valid and robust knowledge in physics.[2,3] Rather than depending on the instructor as the source of knowledge, in the CPU classroom students develop, test and modify their own ideas through experimentation and discussion with their peers. This does not mean there is no organized structure to the classroom. Indeed, there is a carefully designed sequence of activities and a pedagogy that promotes and values extensive intra-group and whole-class discussion. Students also make extensive use of the computer. However, the students' own ideas, supported through experimental evidence, become the standard of authority. In this paper, I will briefly describe the CPU pedagogy, and then focus on the myriad ways that the computer supports learning.

The CPU Classroom

The CPU Project has developed independent units in Light and Color, Static Electricity and Magnetism, Current Electricity, Force and Motion, Waves and Sound, the Nature of Matter and a special skills-oriented unit called Underpinnings.[4,5] The full implementation of each unit requires between 35 and 50 hours of classroom time, so in our one-semester course we either work through two units completely, or portions of three of the units. Each section of our course has 30 students and meets twice-weekly for approximately 2 1/4 hours in a single room with laboratory tables and computers. Although part of the time students are involved in whole-class discussions, the majority of class time is spent with students working in small groups of three or four, performing experiments and recording results on the computer.

Each CPU unit is divided into cycles, each intended to support students' construction of a relevant model or component of a model. Each cycle is divided into three phases: Elicitation, Development and Application.[6] A pedagogical flow diagram is shown in Figure 1.

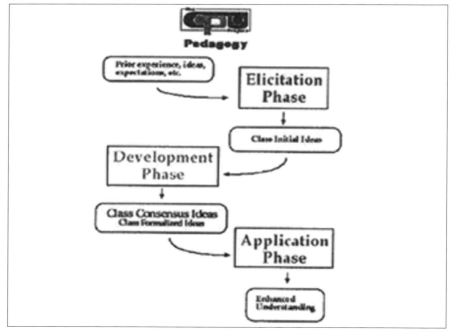

Figure 1

The **elicitation phase** engages students in an extensive and robust discussion centered around some interesting phenomenon. The students are usually asked to make predictions, explain their predictions based on prior knowledge, observe the outcome of the experiment and then to suggest ways of making sense of the outcome, which is often a surprise to many students. Each group shares its initial ideas with the rest of the class using white boards. Figure 2 shows a student presenting her group's idea for how a lens forms an image and her group's prediction for what would happen to the image when half the lens is covered. This elicitation activity is from the third Cycle of the Light and Color Unit. The purpose of this activity is not to make judgments on which ideas suggested by the class are the most "correct" ones, but instead to open up important issues and ideas that make sense to at least some of the students in the class and can serve as focal points of further inquiry.

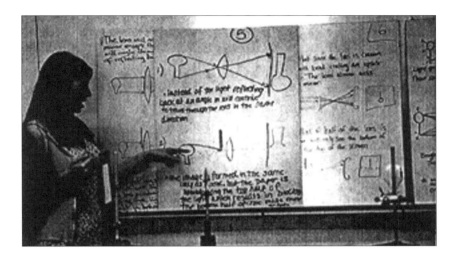

Figure 2

In the **development phase**, students work in small groups (more or less independently), testing the class initial ideas in a wide variety of experimental (hands-on) contexts. They record their observations, ideas and explanations on the computer with special software and use computer simulators to receive both phenomenological and model-based feedback. The sequencing of activities within the units is designed to challenge the common student ideas that are described in the vast literature on research in student understanding. As students go through the development phase, they modify some of their initial ideas, cast some aside as not being useful and invent new ideas. They use electronic *idea journals* to describe and keep a history of their evolving ideas and to record the experimental evidence that supports or refutes these ideas. At the end of the development phase each group is responsible for proposing to the whole class a set of candidate ideas that it believes will best explain the phenomena encountered throughout the cycle and which it can support with evidence. Figure 3 shows a white board constructed by a group at the end of Cycle 3 of the Light and Color Unit. The instructor then leads a whole-class discussion in which all the groups' candidate ideas are consolidated into a set of evidence-supported class consensus ideas. If the development cycle activities were well designed and robust, these ideas should be closely aligned with target ideas for the cycle, except they are phrased in terms of the students' own words, devoid of jargon, convention and formalism. At this point in the pedagogy, the instructor (speaking as a representative of the scientific community) then introduces appropriate technical jargon, conventions and formalism, where appropriate. This formalization of the ideas is important because students would certainly come across the more conventional phrasing of these ideas if they read textbooks, search for information on the World Wide Web or discuss physics with students from other classes. However, it is crucial to this pedagogy that the conceptual seeds of these formalized ideas are generated by the students themselves. The students should come to see that the powerful ideas in science are inventions of the human mind and not dictums from authority.

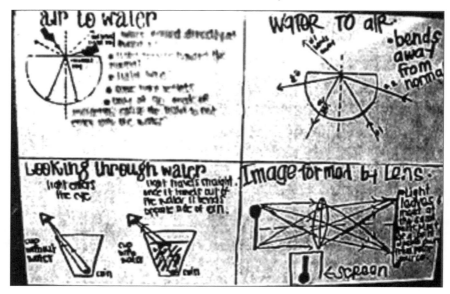

Figure 3

The purpose of the **application phase** is to provide students with myriad opportunities to see the fruitfulness (and perhaps limitations) of the class consensus ideas by applying them in a wide variety of new and interesting contexts. Whereas the development phase engages all students in a carefully structured sequence of activities to ensure they all have a common experiential basis for developing shared (consensus) ideas, the application phase allows students wider latitude to explore their own questions, using available apparatus and the simulators.

The entire learning process is supported by computer-based activity documents and electronic journals, and a set of powerful pedagogically oriented physics simulators.[3] The computer itself, and the simulators in particular, support the learning process in this classroom in several ways, including:

1. Computer facilitates meaningful collaborative discourse among group members
2. Computer helps students document and monitor their (evolving) ideas

3. Simulator enables students to extend hands-on experiments and collect additional phenomenological data
4. Simulator provides conceptual evidence enabling students to test their evolving conceptual models
5. Simulator provides multiple representations referring to the same phenomenon

In the following sections we briefly discuss each of these ways and provide some examples.

The Computer Facilitates Meaningful Collaborative Discourse Among Group Members

We typically have three students in a group, and they situate themselves so that the student who will be the group's typist for the day sits in the middle. The other students sit on either side. One of these other students usually collects the necessary experimental apparatus and sets it up on the table, alongside the computer. Most CPU activities begin with a brief statement of the purpose of the activity, then describe an experimental set-up and ask students to make a prediction and to explain their reasoning. Either one member of the group reads the instructions from the computer screen out loud while other group members follow along, or each member reads the instructions quietly. In either case, the fact that all group members are reading the same set of instructions on the computer screen fosters collaboration. Although one student controls the mouse and does the typing, often all members of the group are looking at the screen and suggesting words to say. It is not uncommon for all three group members to contribute to the response as it is being typed. In that sense, the response is truly a collaborative one.

Often students gesture and point to various objects on the screen (both text and graphics) as a way to clarify what they are saying. The visual objects become a common reference for student discussions. See Figure 4.

Figure 4

The Computer Helps Students Document and Monitor Their Ideas

The purpose of most activities in the development phase of the CPU pedagogy is to enable students to collect evidence in support of ideas. The last step of each activity directs students to open their Idea Journal and to add to or modify their existing ideas, or to develop new ones, and to include supporting evidence. Sometimes they will paste a picture from the activity itself as part of their supporting evidence. Figure 5 shows an example of one group's Idea Journal for the Cycle of the Light and Color Unit that focuses on reflection and mirror images. In this case, on Feb. 22 the students did an experiment in a darkened room, which involved using a flashlight to illuminate a collection of tiny mirrors and tiny pieces of white paper sitting horizontally on a table. Although the illuminated pieces of paper could be seen from any direction, these students did not interpret this observation as indicating light actually reflected from the paper. On Feb. 24, however, they performed another experiment that provided strong evidence to support the existence of this reflection, and they noted that in their Idea Journal.

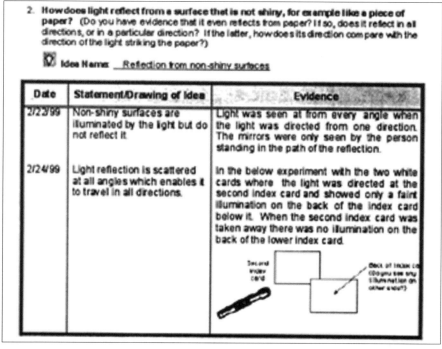

Figure 5

Simulator Enables Students to Extend Hands-On Experiments and Collect Additional Phenomenological Data

In most activities, students collect data by performing experiments with hands-on apparatus. They often follow this by using one of the CPU computer simulators to extend their observations and to collect additional evidence. One advantage of using a simulator is that they can then take a snapshot of the simulator results and copy it into their own activity document. Another advantage is that the additional evidence collected from the simulator can be very helpful in promoting the development of particular ideas. As an illustration, we will describe two related examples, one from the Light and Color Unit and the other from the Waves and Sound Unit.

One of the main ideas we aim to promote in the Light and Color Unit is that an extended source (like a line source) can be thought of as a sequence of closely spaced point sources. (This idea is very useful in understanding the image formation process.) We promote this idea by providing multiple contexts (shadows, pinholes, mirror images and light images) where students can infer that an extended source behaves like a sequence of closely spaced point sources. Figures 6, 7 and 8 illustrate how this is done in the context of shadows. Figure 6 shows the apparatus for an experiment students perform with two point sources, a square-shaped blocker and a screen. (They use Mini-Maglites™ as point sources.) After then considering what happens to the shadow as more and more point sources are added, the students next consider an extended linear source. See Figure 7. (The pictures of the complex shadows on the screen are actually screen-shots from the CPU Shadows and Pinholes simulator.) Figure 8 shows the portion of the activity document where students are explicitly asked to think about the relationship between the continuous source and the sequence of point sources. The figure also includes the responses of a particular group. (Students always type their responses into blank cells with a thick blue border. That way, when they print out the activity documents at the end of each period, they can easily distinguish their own typed responses or drawings from those provided in the activity itself.

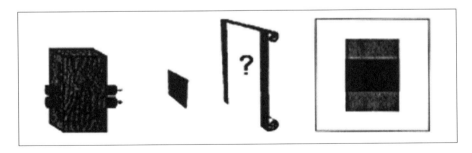

Figure 6

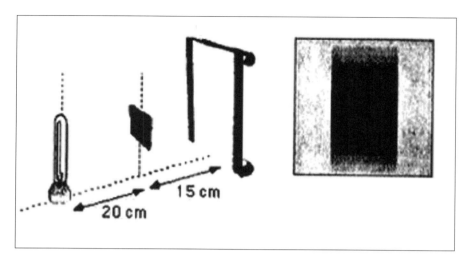

Figure 7

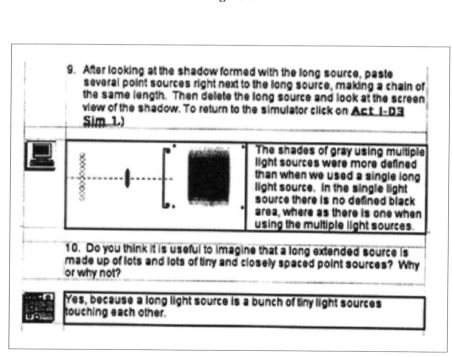

Figure 8

The idea that the behavior of a continuous source can be approximated by a sequence of point sources can also be promoted in the Waves and Sound Unit. In this unit, the teacher can demonstrate the use of a ripple tank as a means to observe the behavior of water waves in simple situations, then have the students perform a series of experiments with a simulated ripple tank. The CPU Ripple Lab simulator allows students to set up several wave tanks simultaneously on the screen and explore how changing one or more parameters changes the resulting wave pattern. Figure 9 shows a snapshot from a single computer screen where four different wave tanks have been arranged simultaneously. The purpose of this sequence is to suggest how the wave pattern from a line source can be approximated as the wave pattern from a sequence of point sources.

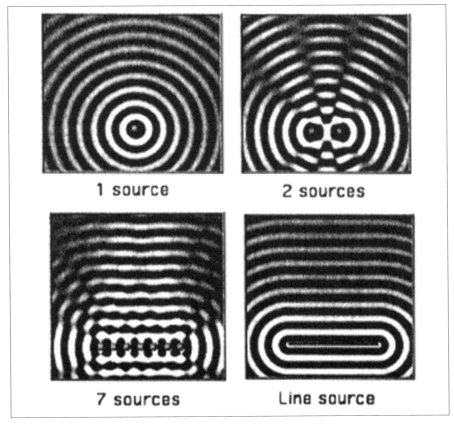

Figure 9

Simulator Provides Conceptual Evidence Enabling Students to Test Their Evolving Conceptual Models

In addition to providing students with additional phenomenological evidence (as described in the two examples above), the computer simulators can also provide conceptual evidence.[7] In this case the computer simulator has a built-in model that could be used in explaining a phenomenon. When students are asked to explain the phenomenon by constructing some type of diagrammatic representation (model), they can obtain evidence from the simulator to compare with their own construction. (To facilitate this comparison, they copy snapshots of the simulator representations into their own document.) Below we provide examples from the CPU Units on Light and Color, Static Electricity and Magnetism, and the Nature of Matter.

In the Light and Color Unit students are often asked to construct light-ray diagrams to explain how light behaves in particular situations. One of the tools available in the Light and Color simulators is a light-ray spray. Students can construct a set-up of simulated apparatus, then drag out a light ray spray from a point on a light source, and the simulator will show how the light rays propagate through the optical system. The simulators also allow a great deal of flexibility in how students can manipulate the light-ray spray. Figure 10 shows an example from the Lens simulator. Here the student has placed a complex source, lens and screen in the set-up window of the simulator. The sequence of screen shots shows what happens when the student drags the origin of the light spray from the top of the complex source towards the bottom. As this is done, the corresponding image point is mapped out on the screen. (The simulator also calculates the distribution of light intensity on the screen and provides a view of what would be seen on the front of the screen.)

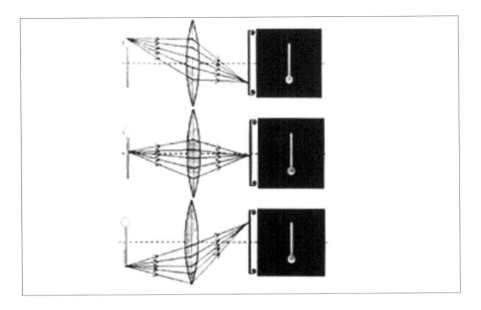

Figure 10

The static electricity simulators provide another example of how students can obtain conceptual evidence. During the Static Electricity and Magnetism Unit, students gather evidence to support the idea that when certain dissimilar objects (like Styrofoam™ and Plexiglas™ plates) are rubbed together, the rubbed surfaces of the two objects are affected differently; that is, when each of these rubbed surfaces are brought near a third rubbed surface, different attraction and repulsion effects are observed. The simulator uses a simple coloring model to support these inferences. When appropriate objects are rubbed together in the simulator, the rubbed surfaces are colored either red or blue, and the thickness of the colored layer depends on the amount of rubbing. (Later in the unit, the red coloring will be associated with excess positive charge, the blue coloring with excess negative charge and the thickness of the layer with the amount of excess surface charge.) Figure 11 shows a snapshot from a simulator that represents what happens when a "red-charged" insulator (the object on the right) is brought close to a "neutral" conductor. The object on the left simulates the behavior of a soda-can electroscope,[8] which has a piece of tinsel hanging from its opposite side. The simulator represents the polarization effect by coloring the opposite sides of the conductor blue and red, and showing the "tinsel" sticking out.

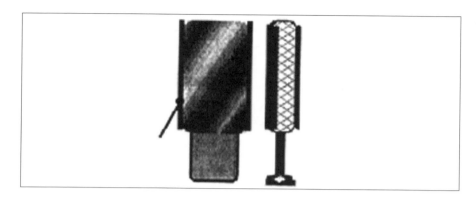

Figure 11

In one of the activities in the unit, the students bring a "charged" plastic straw (having been rubbed with wool) near one side of the soda can, and observe the tinsel sticking out on the other side. Their task is to try to make sense of this observation and to explain it in terms of the red and blue coloring model. By using the simulator to observe dynamically the coloring taking place when the simulated insulator is dragged near to the simulated electroscope, the students are able to develop a reasonable initial explanation for the polarization process. Figure 12 shows a student using her fingers to describe how "charges" in the conductor are being influenced by the "charged" straw. After developing a satisfactory explanation (from the students' point of view), they repeat the explanation with the apparatus (see Figure 13).

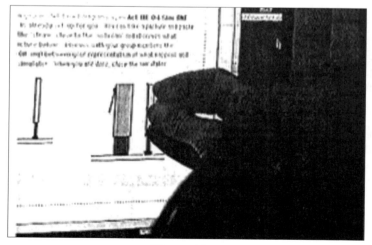

Figure 12

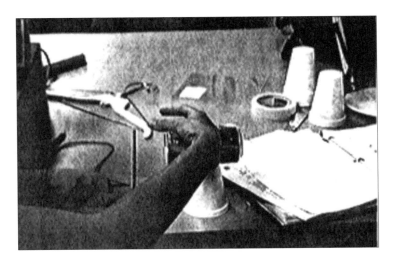

Figure 13

The final example in this section shows how students can obtain both phenomenological and conceptual evidence from one of simulators used in the Nature of Matter Unit. Figure 14 shows a screen shot from the CPU Ideal Gas simulator. On the left, a student can create an experiment by leaking gas into a container that has a movable top wall. Meters can be placed in the container to measure various macroscopic quantities, in this case, density, pressure, temperature and volume. The student can then use a special microscopic viewer tool to obtain a model view of the particles in a tiny volume inside the container. Special meters can record the values of various quantities inside this tiny volume (for example, number of particles, speed, etc.). Students can then graph any of the macroscopic or microscopic quantities that are measured with meters.

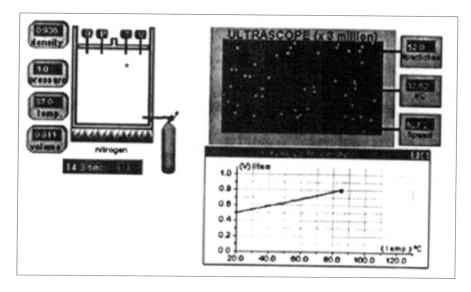

Figure 14

Simulator Provides Multiple Representations Referring to the Same Phenomenon

Many of the simulators have multiple representations of the same phenomenon, and this can aid students' understanding. For example, in the Current Electricity Unit students work with batteries and bulbs to construct a model to explain the behavior of circuits. They often then use the Current Electricity simulator to extend their observations. Figure 15 is a snapshot from the simulator and shows multiple representations of the current concept. This is a circuit with three identical (1.5-volt) batteries and two different bulbs. A compass and ammeter have been added. A separate compass window shows the compass deflection, and the ammeter provides a direct digital readout. The simulator can also represent current in terms of current arrows (whose length is proportional to the value of the current) and current numbers appearing alongside bulbs (whose magnitude is proportional to the current in the bulb). A yellow disk centered on each bulb symbol represents the brightness. (Actually, the area of the disk is proportional to the power dissipated in the bulb.)

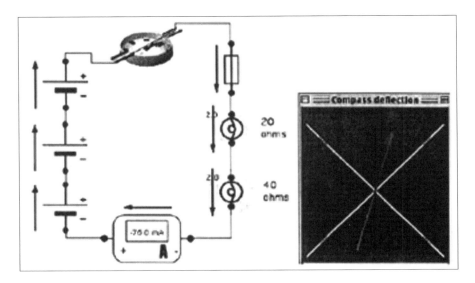

Figure 15

The Force and Motion Unit makes extensive use of the MBL sonar detector (and the force detector) for students to collect data.[9] Figure 16 shows sample data for a fan cart speeding up uniformly along a track.

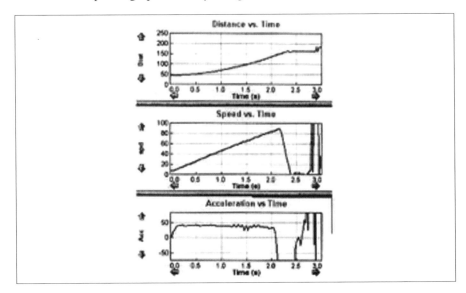

Figure 16

After collecting data with the MBL apparatus, students can repeat the experiment with the CPU Force and Motion simulator. In this case, along with the graphical representations, the simulator can also represent the motion with a strobe diagram and kinematic (velocity and acceleration) arrows. See Figure 17. With the simulator, the students can also easily change the initial speed or acceleration and see the results immediately in multiple representations.

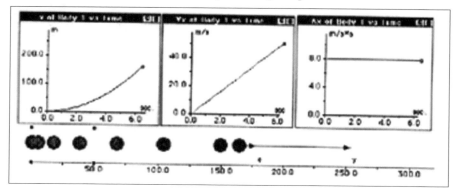

Figure 17

Conclusion

The CPU Pedagogy and supporting curriculum materials were designed to help students develop robust and valid ideas in physics. When students have the main responsibility for developing their own ideas, they need support. In this paper, we have briefly described and illustrated with examples several ways that the computer can provide this support. First, the computer provides a common platform for students in a group to discuss and record their predictions, observations and explanations. Often what appears on the computer screen is not so much the idea of a single student, but instead represents the collaborative effort of the group. Second, the computer can be used as a tool for students to record their evolving ideas. Third, specially designed computer simulations can provide phenomenological evidence that students can use to extend their observations with hands-on equipment. Fourth, the simulations can also provide conceptual evidence that students can use to compare directly with their own conceptual models of the phenomena. Finally, the computer simulations can be used to provide multiple representations of the same phenomena, which can help extend students' understanding.

Acknowledgments

The work described in this paper was supported by the National Science Foundation grant ESI-9454341. Many of the ideas discussed in this paper were developed in collaboration with Andy Johnson and Valerie Otero, who were both graduate students working with me on the CPU Project. Andy is now at Black Hills State University in Spearfish, SD, and Valerie will be at the University of Colorado, Boulder, beginning in January 2001.

Notes and References

1. The CPU Project has been supported by National Science Foundation Grant ESI-9454341. Information about the project is available on its web page, http://cpuproject.sdsu.edu.
2. Goldberg, Fred, "How can computer technology by used to promote learning and communication among physics teachers?," in Rigden, J. (Ed.) *Proceedings of the International Conference on Undergraduate Physics Education Volume I*, American Institute of Physics, 1997.
3. Goldberg, Fred, "Constructing physics understanding in a computer-supported learning environment," in Rigden, J. (Ed.) *Proceedings of the International Conference on Undergraduate Physics Education Volume II*, American Institute of Physics, 1997.
4. The CPU curriculum materials and software were developed by a large team of physics educators. Principal authors and designers included Fred Goldberg (director), Patricia Heller (co-director), Sharon Bendall, Robert Morse, Jim Minstrell, Paul Hickman, Jennifer Hickman, Andy Johnson, Valerie Otero, Laura McCullough, Sandra Grindle, Roy McCullough, Jodi McCullough, Michael McKean, Arni McKinley and Joseph Faletti. The software had been developed in collaboration with Physicon Ltd. (Russia), a member of OpenTeach(c) Group.
5. The CPU Simulation Software and the CPU Curriculum Units are available from The Learning Team, 800/793-TEAM, sales@learningteam.org, or http://www.learningteam.org.
6. This approach is an extended modification of the Learning Cycle, developed by Robert Karplus and others as part of the Science Curriculum Improvement Study (SCIS) of the 1960s.
7. Otero, Valerie, Andy Johnson and Fred Goldberg (in press), "How does the computer facilitate the development of physics knowledge by prospective elementary teachers?," Journal of Education.
8. Morse, Robert A., *Teaching About Electrostatics, AAPT/PTRA Workshop Manual*, American Association of Physics Teachers, College Park, MD, 1992.
9. MBL equipment and software available from PASCO scientific (800/772-8700) or Vernier Software (info@vernier.com).

Preparing Teachers to Teach Physics and Physical Science by Inquiry

Lillian C. McDermott and Peter S. Shaffer
University of Washington

Introduction

The Physics Education Group at the University of Washington is deeply involved in preparing K–12 teachers to teach physics and physical science by inquiry. During the academic year, the Department offers special courses for preservice (prospective) teachers. During the summer, the group conducts a six-week, intensive NSF Summer Institute for Inservice Teachers. The group also designs and helps conduct local short-term workshops for teachers. This paper is a distillation of more than 25 years of experience in working with K–12 teachers.[1]

Teacher preparation has been an integral part of our group's comprehensive program in research, curriculum development, and instruction. Research by our group focuses on investigations of student understanding in physics. The results are used to guide the design of instructional materials for various student populations at the introductory level and beyond. We have drawn on our research findings and teaching experience in developing *Physics by Inquiry*.[2] This self-contained, laboratory-based curriculum is designed for use in university courses to prepare K–12 teachers to teach physics and physical science effectively. Ongoing assesment of the instructional materials takes place both at our university and at pilot sites.

Need for Special Physics Courses for Teachers

Most science departments, including physics, do not take into account the needs of prospective elementary and middle school teachers. These students often lack the prerequisites for even the standard introductory courses, especially in the physical sciences. They are unlikely to pursue the study of any science in depth because the vertical structure of the subject matter requires progression through a prescribed sequence of courses. In physics, in particular, the need for mathematical facility in the standard courses effectively excludes those planning to teach below the high-school level. The only courses generally available are almost entirely descriptive. A great deal of material is presented, for which most preservice and inservice teachers (as well as other students) have neither the background nor the time to absorb. Such courses often reinforce a tendency to perceive physics as an inert body of information to be memorized, not as an active process of inquiry in which teachers and students can participate.

Many university faculty seem to believe that the effectiveness of a high school teacher depends on the number and rigor of courses taken in the discipline. This attitude seems to prevail in most physics departments. Accordingly, the usual practice is to offer the standard department courses to future high school physics teachers (and sometimes to middle school teachers). Although the content of the high school physics curriculum is closely matched to the introductory university course, this course is not adequate preparation for teaching the same material in high school. The breadth of topics covered allows little time for acquiring a sound grasp of the underlying concepts. The routine problem solving that characterizes most introductory courses does not help teachers develop the reasoning ability necessary for handling the unanticipated questions that are likely to arise in a classroom. The laboratory courses offered by most physics departments also do not address the needs of teachers. Often the equipment is not available in high schools, and no provision is made for showing teachers how to plan laboratory experiences that utilize simple apparatus. A more serious shortcoming is that experiments are mostly limited to the verification of known principles. Students have little opportunity to start from their observations and go through the reasoning involved in formulating these principles. As a result, it is possible to complete a laboratory course without confronting conceptual issues or understanding the scientific process.

For those students who progress beyond the first year of university physics, advanced courses are of little direct help in teaching. The abstract formalism that characterizes upper-division courses is not of immediate use in the precollege classroom. Sometimes, in the belief that teachers need to update their knowledge, university faculty may offer courses on contemporary physics for preservice or inservice teachers. Such courses are of limited utility. The information may be motivational but does not help the teachers recognize the distinction between a memorized description and substantive understanding of a topic. Although work beyond the introductory level may help teachers deepen their understanding of physics, no guidance is provided about how to make appropriate use of this knowledge in teaching high school students.

A well-prepared teacher of physics or physical science should have, in addition to a strong command of the subject matter, knowledge of the difficulties it presents to students. Traditional courses in physics do not provide this kind of preparation. They also have another major shortcoming. Teachers tend to teach as they were taught. If they were taught through lecture, they are likely to lecture, even if this type of instruction is inappropriate for their students. Many teachers cannot, on their own, separate the physics they have learned from the way in which it was presented to them.

Development of Physics Courses for Teachers

The emphasis in courses for teachers should be on the development of deep understanding of topics included in the K–12 curriculum. Teachers should study each topic in a way that is consistent with how they are expected to teach that material. The intellectual objectives and instructional approach in courses for teachers should be mutually reinforcing.

Intellectual Objectives

Teachers need the time and guidance to learn basic physics in depth, beyond what is possible in standard courses. They should be given the opportunity to examine the nature of the subject matter, to understand not only what we know, but on what evidence and through what lines of reasoning we have come to this knowledge. A sound conceptual understanding of basic physics and capability in scientific reasoning provide a firmer foundation for effective teaching than superficial learning of more advanced material.

It is critical that teachers be able to do the qualitative and quantitative reasoning that underlie the development and application of concepts. Instruction for teachers should cultivate scientific reasoning skills, which tend to be overlooked in traditional courses. It has been demonstrated, for example, that university students enrolled in standard physics courses often cannot reason with ratios and proportions.[3] Proportional reasoning is obviously a critically important skill for high school science teachers, but it is also essential for elementary and middle school teachers who are expected to teach science units that involve concepts such as density and speed.

Although high school teachers must be able to solve textbook problems, the emphasis in a course for teachers should not be on mathematical manipulation. As necessary as quantitative skills are, ability in qualitative reasoning is even more critical. Teachers need to recognize that success on numerical problems is not a reliable measure of conceptual development. They should be given a great deal of experience with questions that require careful reasoning and explanations.

It is also necessary for teachers to develop skill in using and interpreting formal representations, such as graphs, diagrams and equations. To be able to make the formalism of physics meaningful to students, teachers must be adept at relating different representations to one another, to physical concepts and to real-world phenomena.

An understanding of the nature of science should be an important objective in a course for teachers. Teachers at all grade levels must be able to distinguish observations from inferences and to do the reasoning necessary to proceed from observations and assumptions to logically valid conclusions. They must understand what is considered evidence in science, what is meant by an explanation and what the difference is between naming and explaining. The scientific process can only be taught through direct experience. An effective way of providing such experience is to give teachers the opportunity to construct a conceptual model from their own observations. They should go step-by-step through the process of making observations, drawing inferences, identifying assumptions, formulating, testing and modifying hypotheses. The intellectual challenge of applying a model that they themselves have built (albeit with guidance) to predict and explain progressively more complex phenomena can help teachers deepen their own understanding of the evolving nature, use and limitations of a scientific model. We have also found that successfully constructing a model through their own efforts helps convince teachers (and other university students) that reasoning based on a coherent conceptual framework is a far more powerful approach to problem solving than rote substitution of numbers in memorized formulas.

The instructional objectives discussed above are, in principle, equally appropriate for the general student population. However, teachers have additional requirements that special physics courses should address. For example, teachers need to develop skill in formulating and applying operational definitions. To be able to help students distinguish between related but different concepts (e.g., velocity and acceleration), teachers must be able to describe precisely and unambiguously how the concepts differ and how they are related. Teachers also need to be given the opportunity to confront and resolve their conceptual and reasoning difficulties, not only to improve their own learning but to become aware of the difficulties that their students will have.

Courses for teachers should help develop the critical judgment necessary for making sound choices on issues that can indirectly affect the quality of instruction in the schools. For example, teachers must learn to discriminate between meaningful and trivial learning objectives. When instruction is driven by a list of objectives that are easy to achieve and measure, there is danger that only shallow learning, such as memorization of factual information, will take place. Teachers also need to develop criteria for evaluating instructional materials, such as science kits, textbooks, laboratory equipment and computer software. They should be able to identify strengths and weaknesses in school science programs. Through service on district committees, individual teachers can often have an impact that extends beyond their own classrooms. Aggressive advertising and an attractive presentation often interfere with objective appraisal of the intellectual content of printed materials or computer software. Teachers should learn to resist the temptation of an appealing program format when there are serious flaws in physics. A poor curriculum decision can easily deplete the small budget most school districts have for science without resulting in an improvement in instruction.

Instructional Approach

If the ability to teach by inquiry is a goal of instruction, teachers need to work through a substantial amount of content in a way that reflects this spirit. The instructional approach in our courses for teachers can be characterized as guided inquiry.

Science instruction for young students is known to be more effective when concrete experience establishes the basis for the construction of scientific concepts.[4] We and others have found that the same is true for adults, especially when they encounter a new topic or a different treatment of a familiar topic. Therefore, instruction for prospective and practicing teachers should be laboratory-based. However, "hands-on" is not enough. Unstructured activities do not help students construct a coherent conceptual framework. Carefully sequenced questions are needed to help them think critically about what they observe and what they can infer. When students work together in small groups, guided by well-organized instructional materials, they can also learn from one another.

Whether intended or not, teaching methods are learned by example. The common tendency to teach physics from the top down, and to teach by telling, runs counter to the way precollege (and many university) students learn best. The instructor in a course for teachers should not transmit information by lecturing, but neither should he or she take a passive role. The instructor should assume responsibility for student learning at a level that exceeds delivery of content and evaluation of performance. Active leadership is essential, but in ways that differ markedly from the traditional mode. This approach, which can be greatly facilitated by a well-designed curriculum, is characterized below in general terms and illustrated in the next section in the context of specific subject matter. Other examples are given in published articles.[5]

The instructional materials used in a course for teachers should be consistent with those used in K–12 science programs, but the curriculum should not be identical. Teachers must have a deeper conceptual understanding than their students are expected to achieve. They need to be able to set learning objectives that are both intellectually meaningful for the topic under study and developmentally appropriate for the students.

The study of a new topic should begin with open-ended investigation in the laboratory, through which students can become familiar with the phenomena of interest. Instead of introducing new concepts or principles by definitions and assertions, the instructor should set up situations that suggest the need for new concepts or the utility of new principles. By providing such motivation, the instructor can begin to demonstrate that concept formation requires students to become mentally engaged. Generalization and abstraction should follow, not precede, specific instances in which the concept or principle may apply. Once a concept has been developed, the instructor should present new situations in which the concept is applicable but may need to be modified. This process of gradually refining a concept can help develop an appreciation of the successive stages that are involved in developing a sound conceptual understanding.

As students work through the curriculum, the instructor should pose questions designed to help them to think critically about the subject matter and to ask questions on their own. The appropriate response of the instructor to most questions is not a direct answer but another question that can help guide the students through the reasoning necessary to arrive at their own answers. Questions and comments by the instructor should be followed by long pauses in which the temptation for additional remarks is consciously resisted. Findings from research indicate that the quality of student response to questions increases significantly with an increase in "wait time," the time the instructor waits without comment after asking a question.[6]

As mentioned earlier, a course for teachers should develop an awareness of common student difficulties. Some are at such a fundamental level that, unless they are effectively addressed, meaningful learning of related content is not possible. Serious difficulties cannot be overcome through listening to lectures, reading textbooks, participating in class discussions, or consulting references. Like all students, teachers need to work through the material and have the opportunity to make their own mistakes. When difficulties are described in words, teachers may perceive them as trivial. Yet we know that often these same teachers, when confronted with unanticipated situations, will make the same errors as students. As the opportunity arises during the course, the instructor should illustrate instructional strategies that have proved effective in addressing specific difficulties. If possible, the discussion of a specific strategy should occur only after it has been used in response to an error. Teachers are much more likely to appreciate important nuances through an actual example than through a hypothetical discussion. Without specific illustrations, it is difficult for teachers to envision how to translate a general pedagogical approach into a specific strategy that they can use in the classroom.

The experience of working through a body of material step-by-step can help teachers identify the difficulties their students may have. A considerable amount of research has been done on difficulties common to students at all levels (K–20) of physics education. Faculty who teach teachers should be familiar with this resource and be able to refer them to relevant articles. Teachers who understand both the subject matter and the difficulties it poses for students are likely to be more effective than those who know only the content.

Because it is critical that teachers be able to communicate clearly, group discussions and writing assignments should play an important role in a physics course for teachers. Providing multiple opportunities for teachers to reflect upon and to describe their own conceptual development can enhance both their knowledge of physics and their ability to formulate the kinds of questions that can help their students deepen their understanding.

Illustrative Example: Electric Circuits

We can illustrate the research basis and instructional approach that guide the development of our courses for teachers with a specific example. The topic of electric circuits is included in almost all K–12 standards–based science curricula.

1. Investigation of conceptual understanding

Research by our group on student understanding of electric circuits has extended over a period of many years.[8] Since the results are well known by now, only a brief discussion is presented here. The question shown in Figure 1 has been given to many different groups of students and teachers. The question asks for a ranking by brightness of the five identical bulbs in the circuits shown and to explain their reasoning. The batteries are ideal. The correct response is A = D = E > B = C.

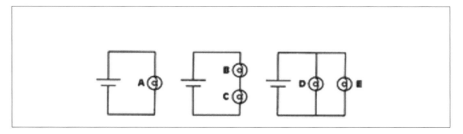

Figure 1. The five bulbs are identical and the batteries are identical and ideal. Rank the five bulbs from brightest to dimmest. Explain your reasoning.

This question has been administered to more than 1,000 students in introductory calculus-based physics. Whether before or after standard instruction in lecture and laboratory, student performance has been essentially the same. Only about 15% of the students have responded correctly. The same question has produced similar results when administered to high school physics teachers and to university faculty in other sciences and mathematics. Analysis of the responses revealed the lack of a conceptual model for a simple electric circuit. Reliance on the rote use of inappropriate formulas was common.

We have also had opportunities to pose this question to graduate students in our Department's Physics PhD program. Approximately 70% have given the correct response. Since these students are Teaching Assistants in our introductory physics courses, we have felt it important to help them develop a coherent conceptual model through an abbreviated version of the approach that we use in our courses for teachers.

2. Instruction by guided inquiry

To prepare teachers to teach the topic of electric circuits by inquiry, we engage them in the step-by-step process of constructing a qualitative model that they can use to predict and explain the behavior of simple circuits that consist of batteries and bulbs.[9] Mathematics is not necessary. Qualitative reasoning is sufficient. Specific difficulties that have been identified through research are addressed during the development of the model. Two of the most common are the apparent belief that the battery is a constant current source and that current is "used up" in a circuit.

Students are guided through carefully sequenced activities and questions to make observations that they can use as the basis for their model. The students begin the process of model-building by trying to light a small bulb with a battery and a single wire. They develop an operational definition for the concept of a complete circuit. Exploring the effect of adding additional bulbs and wires to the circuit, they find that their observations are consistent with the assumptions that a current exists in a complete circuit and the relative brightness of identical bulbs indicates the relative magnitude of the current. As the students conduct further experiments—some suggested, some of their own devising—they find that the brightness of individual bulbs depends both on how many are in the circuit and on how they are connected to the battery and to one another. The students are led to construct the concept of electrical resistance and find that they can predict the behavior of many, but not all, circuits of identical bulbs. They recognize the need to extend their model beyond the concepts of current and resistance to include the concept of voltage (which will later be refined to potential difference). As bulbs of different resistance and additional batteries are added, the students find that they need additional concepts to account for the behavior of more complicated circuits. They are guided in developing more complex concepts, such as electrical power and energy. Through deductive and inductive reasoning, the students construct a model that can account for relative brightness in any circuit consisting of batteries and bulbs.

It is important that teachers be asked to synthesize what they have learned, to reflect on how their understanding of a particular topic has evolved and to try to identify the critical issues that need to be addressed for meaningful learning to occur. As they progress in their investigation of electric circuits, the teachers are given many opportunities to express their ideas in writing.

The instructional approach that has been illustrated in the context of electric circuits has proved effective with teachers at all levels from elementary through high school. The process of hypothesizing, testing, extending and refining a conceptual model to the point that it can be used to predict and explain a range of phenomena is the heart of the scientific method. It is a process that must be experienced to be understood.

3. Assessment of effectiveness

Although many of the teachers in our preservice and inservice courses have had considerably less preparation in physics than those in the standard introductory courses, their performance on qualitative questions has been consistently better. The question shown in Figure 2 provides a good example of what elementary teachers without a strong mathematical background, but with a good conceptual understanding, can do. The students are asked to rank the five bulbs in the circuit according to brightness. Reasoning on the basis of a model based on the concepts of current and resistance, most of the elementary and middle school teachers who have been through our courses for teachers predict correctly that $E > A = B > C = D$. This question is beyond the capability of most students who have had standard instruction.

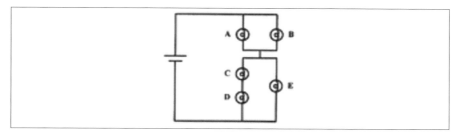

Figure 2. The five bulbs are identical and the batteries are identical and ideal. Rank the five bulbs from brightest to dimmest. Explain your reasoning.

In Figure 3 is a circuit that has been used as a post-test in our NSF Summer Institute for Inservice Teachers. After working through a significant portion of the module on electric circuits in *Physics by Inquiry*, virtually all of the teachers (N = 100) were able to predict and to explain, on the basis of the qualitative model that they had developed, that $A > E > B > C = D$. This represents an improvement over their performance of 15% correct on the pretest in Figure 1.

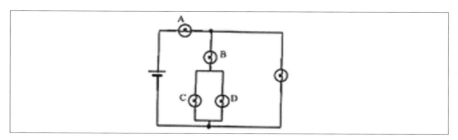

Figure 3. The five bulbs are identical and the batteries are identical and ideal. Rank the five bulbs from brightest to dimmest. Explain your reasoning.

Other evidence about the effectiveness of the approach taken in *Physics by Inquiry* comes from a colleague at the University of Cyprus.[10] (A Greek translation of *Physics by Inquiry* was used.) The performance of two main groups of prospective elementary school teachers was compared. Both groups were taught by instructors who understood the material well. One of the groups (N = 189) had studied electric circuits in the way that has been described. This group consisted of two classes: one had just completed study of the material (N = 102); the other had completed study the previous year (N = 87). The second main group (N = 101) consisted of teachers who had just completed the topic in a course in which constructivist pedagogy was strictly followed (*i.e.*, the students were actively involved in the construction of concepts.) However, instruction in this course was not guided by findings from the type of discipline-based research that has been described. Specific difficulties were not explicitly addressed nor was there the same degree of emphasis on the development of a coherent conceptual model. Two types of post-tests were given: one consisted of free-response questions that asked for explanations of reasoning; the other contained multiple-choice questions taken from DIRECT, a test developed at North Carolina State University.[11]

As can be seen in Figure 4, both classes of students who had studied the material in *Physics by Inquiry* had mean scores greater than 80% on both tests (a result that indicates that retention was very good). In the other main group, performance on the multiple-choice test was slightly above the 40% level. On the free-response test, the mean score was less than 20%. Courses in which educational methodology is emphasized without sufficient emphasis on concept development seem to be no more effective than standard physics instruction.

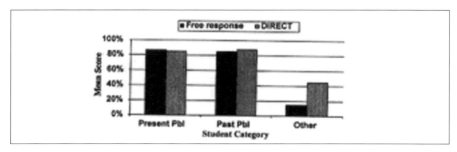

Figure 4. Student performance on free response and on multiple-choice questions on a post-instruction survey on electric circuits. The survey was administered to preservice elementary school teachers at the University of Cyprus. Two main groups of students were included in the survey: those who had used Physics by Inquiry (PbI) and those who had not. Some of the students had studied PbI one year before taking the test (Past PbI). All the others (Present PbI and Other) had just completed their study of this topic.

Inadequacy of Teacher's Guides for Preparation in Content
Teachers who do not acquire the necessary background for teaching science in appropriate preservice or inservice courses are forced to rely on short workshops conducted by school districts and on the teacher's guides that come with the student materials. Even in instances when these materials are good, the accompanying information for teachers is usually grossly inadequate. A few quotes from two well-known elementary-school programs can serve as illustrations.

From one K–6 program:[12]
- "It is not necessary to have studied electricity to teach this module."
- "A battery is said to have so many volts; the more volts, the bigger the push to make current flow. The current itself is measured in amperes. For a given circuit (e.g., a battery, wire, and one bulb) the more volts (push) the more amperes (current)."
- "A given battery exerts a certain amount of push to send electrons around a circuit. …The heat comes from the electrons of the current bouncing against the stationary atoms of the metal as they flow. …even though the metal is solid, most of the space in its atoms is empty and the electrons can move very easily."

From another K–6 program:
- "How does electricity flow along a circuit? Like many things in nature, electricity is invisible, but we can see and measure the results of the flow. The battery, or energy source, gives electricity its 'push' through a circuit. This push, or voltage, can be thought of as electrical pressure, and is analogous to water pressure. Electrical pressure is measured in volts."
- "The actual flow of electricity through a circuit is analogous to the flow of water through a hose. The flow of electrical current is measured in amperes."

Implementation of Physics Courses for Teachers

Like other physics departments, ours offers a number of courses that could be taken by prospective K–12 teachers. As has been discussed, however, such courses provide inadequate preparation for elementary, middle, and high school teachers. At the University of Washington, we have implemented the ideas that have been discussed in this paper in two sets of courses: one for elementary and middle school teachers and the other for high school physics teachers and well-prepared middle school science teachers.

Course Structure

In designing the curriculum for the course for elementary school teachers, we have drawn on the content that is taught in the lower grades. The course does not proceed through the traditional physics sequence (kinematics, dynamics, electricity and magnetism, etc.). The curriculum instead focuses on basic topics (mass, volume, density, temperature, electric circuits, etc.). There is a strong emphasis on proportional reasoning, control of variables and the development of other important scientific reasoning skills.

In the course for prospective high school teachers, the emphasis is on the material covered in a high school physics course, which often covers the same content as the standard introductory university course. The experience of students in the introductory course (and more advanced courses) is often limited to memorization of formulas and numerical problem solving. As in the course for elementary and middle school teachers, the instructional approach in the course for high school teachers is very different. The students go through the reasoning involved in the development of each concept. They are guided in synthesizing what they have learned into a coherent conceptual framework.

Administrative Challenges

A number of challenges must be met in implementing a teacher-preparation program in a physics department, especially at a large, research-oriented university. The argument must be made to the department and higher administrative units that the proposed courses are at an intellectual level worthy of the credit offered. We made this case successfully at our university by demonstrating that the demands on the students matched, or exceeded, those of other physics courses at comparable levels.

Other problems that may need to be addressed relate to enrollment. Mass education does not work for teachers. Laboratory-based instruction is necessary. The classes must be small enough to foster interaction among the students and between the students and the instructor. Sometimes, however, the problem is low total enrollment. It is particularly difficult to reach prospective teachers when there is no undergraduate education major. They are hard to identify since they are unlikely to decide, on their own, to take physics. In the course for prospective elementary and middle school teachers, we have dealt with this issue by encouraging participation of students not planning to major in science.

In the course for prospective high school (and well-prepared middle school) teachers, the enrollment problem has been addressed in a way that has proved to be very effective. We recognized that it would be impossible to fill a class with physics majors who plan to teach. Moreover, most high school physics teachers were not physics majors. At best, they may have majored in chemistry or mathematics. Therefore, we actively encourage participation in the course by students majoring in other sciences and in mathematics. The course is open to all students who have taken the standard introductory physics course. The range of preparation in the class varies from this level to students getting a graduate degree in physics. The system works well because the emphasis is not on quantitative problem-solving but on concept development and reasoning ability. In addition to maintaining a steady enrollment throughout the academic year (20 to 30), we have found that having more non-majors than majors forces students to abandon their dependence on formulas and think more deeply about the physics involved.

Conclusion

The separation of instruction in science (which takes place in science courses) from instruction in methodology (which takes place in education courses) decreases the value of both for teachers. Effective use of a particular instructional strategy is often content specific. If teaching methods are not studied in the context in which they are to be implemented, teachers may be unable to identify the elements that are critical. Thus they may not be able to adapt an instructional strategy that has been presented in general terms to specific subject matter or to new situations. Even detailed directions cannot prevent misuse of excellent instructional materials when teachers do not understand either the content or intended method of presentation. Since the type of preparation that teachers need is not available through the standard physics curriculum, a practical alternative is to offer special courses for teachers. The instructors in such courses must have a sound understanding of the subject matter, of the difficulties that it presents to students and of effective instructional strategies for addressing these difficulties.

It is important for physics faculty to recognize that teachers must be prepared to teach the material at an appropriate level in K–12 classrooms. We have found that teachers often try to implement instructional materials in their classrooms that are very similar to those they have used in their college courses. Through direct experience with the intellectual demands of learning by inquiry, teachers can become better equipped to meet the challenge of matching their instruction to the developmental level of their students.

Our experience indicates that it is not easy to develop good inquiry-oriented instructional materials. Therefore, unless faculty are prepared to devote a great deal of effort over an extended period to the development of a course for teachers, they should take advantage of already existing instructional materials that have been carefully designed and thoroughly tested with teachers. We have found that the sense of empowerment that results when teachers have developed a sound conceptual understanding of the science content that they are expected to teach greatly increases their confidence in their ability to deal with unexpected situations in the classroom.

Acknowledgments

This paper reflects the cumulative experience of many members of the Physics Education Group, past and present. In particular, we would like to acknowledge Paula R. L. Heron and Stamatis Vokos (faculty colleagues in our group) and Lezlie S. DeWater and Donna L. Messina (K–12 teachers in our group) for their substantive contributions to our current teacher preparation program. Special recognition is due to C. P. Constantinou of the University of Cyprus for some of the research results presented in this paper. We appreciate the ongoing support of the National Science Foundation for enabling our group to conduct a coordinated program in research, curriculum development, and instruction.

References

1. This paper builds on other articles by the Physics Education Group on teacher preparation in physics and physical science. For example, see L.C. McDermott, P.S. Shaffer. and C.P. Constantinou, "Preparing teachers to teach physics and physical science by inquiry," *Physics Education* **35** (6) 411 (November 2000); L.C. McDermott and L.S. DeWater, "The need for special science courses for teachers: Two perspectives," an invited chapter in *Inquiring into Inquiry Learning in Teaching and Science*, J. Minstrell and E.H. van Zee, eds., Washington, DC: AAAS (2000, pp. 241-257; and L.C. McDermott, "A perspective on teacher preparation in physics and other sciences: The need for special courses for teachers," *American Journal of Physics*, **58** (8) 734-742 (1990). See also Ref. 8.
2. *Physics by Inquiry, Vols. I and II*, L.C. McDermott and the Physics Education Group at the University of Washington (John Wiley & Sons Inc., New York, NY, 1996).
3. A.B. Arons, *A Guide to Introductory Physics Teaching* (Wiley, New York, 1990), pp. 3-6.
4. See for example, J. Griffith and P. Morrison, "Reflections on a decade of gradeschool science," *Physics Today*, **25** (6), 29-34 (1972); R. Karplus, "Physics for beginners," *Physics Today*, **25** (6), 36-47 (1972); and J. W. Renner, D.G. Stafford, W.J. Coffia, D.H. Kellogg and M.C. Weber, "An evaluation of the Science Curriculum Improvement Study," *School Science and Mathematics*, 73 (4), 291-318 (1973).
5. See, for example, K. Wosilait, P.R.L. Heron, P.S. Shaffer and L.C. McDermott, "Development and assessment of a research-based tutorial on light and shadow," *American Journal of Physics*, **66**, 906-913 (1998); and M.L. Rosenquist and L.C. McDermott, "A conceptual approach to teaching kinematics," *American Journal of Physics*, **55**, 407-415 (1987).
6. Rowe, M.B., "Wait time and rewards as instructional variables, their influence on language, logic, and fate control: Part one–wait time," *Journal of Research in Science Teaching*, **11**, 81-94 (1974).
7. A selection of articles can be found in L.C. McDermott and E.F. Redish, "Resource Letter PER-1: Physics Education Research," *American Journal of Physics*, **67**, 755-767 (1999). Although most of the studies cited in this resource letter refer to students at the university level, similar difficulties have been identified among younger students.
8. McDermott, L.C., and P.S. Shaffer, "Research as a guide for curriculum development: An example from introductory electricity. Part I: Investigation of student understanding," *American Journal of Physics*, **60**, 994-1003 (1992); Erratum to Part I, *American Journal of Physics*, **61**, 81 (1993); and P.S. Shaffer and L.C. McDermott, "Research as a guide for curriculum development: an example from introductory electricity, Part II: Design of instructional strategies," *American Journal of Physics*, **60**, 1003-1013 (1992).
9. The instructional sequence can be found in the *Electric Circuits* module in Volume II of *Physics by Inquiry*. (See Ref. 2.)
10. C.P. Constantinou is a physicist in the Learning in Physics Group, Department of Education, University of Cyprus, Nicosia, Cyprus.
11. DIRECT is a conceptual test of student understanding of DC circuits developed by P.V. Engelhardt and R.J. Beichner, North Carolina State University (unpublished).

12. *Circuits and Pathways, INSIGHTS, Teacher's Guide*, ©1994, Educational Development Center Inc., 55 Chapel St., Newton, MA 02160.

13. *Magnets and Motors, Teacher's Guide*, ©1991, Science and Technology for Children (STC), National Science Resources Center, Smithsonian Institution—National Academy of Sciences, Arts and Industries Building, Room 1201, Washington, DC 20560.

14. See the articles in Ref. 1 and L.C. McDermott, "Combined physics course for future elementary and secondary school teachers," *American Journal of Physics*, **42**, 668-676 (1974); and L.C. McDermott, "Improving high school physics teacher preparation," *The Physics Teacher*, **13**, 523-529 (1975).

Investigating the Role of Physics Departments in the Preparation of K-12 Teachers

John Layman
University of Maryland

My experience at the University of Maryland extends over an eight-year period during which I, with the help of others, designed an introductory physics course for preservice elementary teachers. The course was modeled after the AAPT Powerful Ideas in Physical Science.[1] This became part of the course and program changes associated with the Maryland Collaborative for Teacher Preparation and NSF-sponsored program.[2]

The Nature of the Course
The nature of a physics or physical-science course appropriate for elementary teachers is revealed in some of the statements shared with the students in the course syllabus.

Laboratories
The laboratory activities are the key to the course. Most of the concepts that we deal with will be encountered first in the guise of laboratory activities. Along with your lab group you will observe physical systems, predict their behavior, test your predictions and draw your own conclusions based on your laboratory experiences. You and your lab partners will be the world's most knowledgeable persons in this enterprise and the negotiators of our final understanding of each concept. Your teachers are resources, but they cannot do the understanding for you, nor simply tell you of theirs.

Labgroups
Labgroups will consist of three students, and should become a stable group during the third week. This modest learning community will share much of the responsibility for the personal understanding of all of its members.

Examinations

There will be two two-hour examinations and one final exam. Each will have a written portion as well as a laboratory-activity component, and be worth 150 points.

Exams will be conceptual in nature, comprised largely of essay questions that will require you to draw from your personal laboratory experiences as you articulate and support your understanding of physical concepts. When you are asked to solve a quantitative problem you will also be asked to provide a conceptual explanation. Each examination will contain a laboratory experience to be completed with your own lab group.

You may use your notebook during the exams.

Journals

You will be asked to respond informally to activities, discussions and questions in a journal. This is the place for personal comments, questions and thoughts to be shared with the class, or in a personal interchange with your teachers. The journal can be a critical feature in our learning process. We will offer credit for journal submissions, and evaluate each using a rubric to be described later. Submissions will be electronically supplied on the PHYS 117 listserv, or sent to the instructor.

Written Assignments

Assignments will be given with due dates for each. There will be a combination of laboratory and homework assignments, both of which will contribute to your semester's overall score.

PHYS 117A
SKILLS DISCUSSION
Spring 1997

A number of skills will be developed within our PHYS 117 learning community, for which you have a personal responsibility. From time to time we will relate these to the National Science Education Standards, so you will have a context for your work.

Observation

We will be able to demonstrate with laboratory activities most of the concepts or ideas that we will be dealing with in this introductory physical-science course. The first skill you will need to develop is that of making personal predictions and describing in your own words what you personally observe as activities are carried out. This will sometimes mean utilizing diagrams in support of your words, and using words that are already a natural part of your vocabulary. As our work progresses, you will begin making the transition to words that arise from within the science community, a community within which you are now a full participant.

Scholarly Response

On examinations and in your written work, the first stage of a scholarly response will be your personal skill in describing in your own words your predictions or conjectures, what you have observed, followed by words and explanations that may have been provided by you, your labmates, the professor, the TA, the text or lab guide.

Evidence

The best evidence you can offer is the statement "**I saw it**," not "the book said so" or "Dr. Layman said." This ability becomes your personal responsibility, and our task is to optimize your chance to do this skillfully. One caveat, however, is the statement that may become more clear as the semester progresses, "If I hadn't believed it, I wouldn't have seen it."

Explanations

Explanations for things observed, offered by you, your labmates, your TA, the teacher, the text and from other sources must always be greeted with some skepticism. Our observations on the other hand are more reliable and can always be verified by repeating the observation. We must, however, recognize that although we may all be "observing the same event," we may not all "see" the same thing. When explanations for what we observe involve second-hand information or inferences from the observations, however correct they may turn out to be, we will occasionally use the term "rumor has it," to indicate that we may not yet have full understanding of a concept.

Welcome

Welcome aboard. We will have a grand time honing your personal skills and understandings, and your ability to watch yourself learn. We (meaning you and the instructors) will also learn much from our interactions. All members of a successful learning community participate in "learning."

Note: Students enrolled in a course should have access to the rational behind the course and have an indication that the teaching/learning procedures employed arise from best practice as described or defined in national reports. The remainder of this section of the syllabus is made up of quotations from one national document.

SHAPING THE FUTURE[3]

New Expectations for Undergraduate Education in Science,
Mathematics, Engineering and Technology

A Report on its Review of Undergraduate Education
by
the Advisory Committee to the National Science Foundation
Directorate for Education and Human Resources

Too many students leave Science, Mathematics, Education & Technology (SME&T) courses because they find them dull and unwelcoming. Too many new teachers enter school systems underprepared, without really understanding what science and mathematics are, and lacking the excitement of discovery and the confidence and ability to help children engage SME&T knowledge. Too many graduates go out into the workforce ill prepared to solve real problems in a cooperative way, lacking the skills and motivation to continue learning.[4]

We recommend that[5]:

SME&T faculty: Believe and affirm that every student can learn, and model good practices that increase learning; start with the student's experience, but have high expectations within a supportive climate; and build inquiry, a sense of wonder and the excitement of discovery, plus communication and teamwork, critical thinking and life-long learning skills into learning experiences.

Inquiry—although there is disagreement about the meaning of the term "science literacy" and doubt about whether agreement is possible on a list of facts everyone should know. There is no disagreement that every student should be presented an opportunity to understand what science is, and is not, and to be involved in some way in scientific inquiry, not just a "hands-on" experience.

VII. SME&T faculty[6]

A. Believe and affirm that every student can learn; recognize that different students may learn in different ways and with differing levels of ability; and create an environment in each class that both challenges and supports.
B. Be familiar with and use the results of professional scholarship on learning and teaching.
C. Build into every course inquiry the processes of science (or mathematics or engineering), a knowledge of what SME&T practitioners do and the excitement of cutting-edge research.
D. Devise and use pedagogy that develops skills for communication, teamwork, critical thinking and lifelong learning in each student.

E. Make methods of assessing student performance consistent with the goals and content of the course.
F. Start with the student's experience; understand that the student may come with significantly incorrect notions; and relate the subject matter to things the student already knows.
G. Build bridges to other departments, seeking ways to reinforce and integrate learning, rather than maintaining artificial barriers.
H. Develop partnerships and collaborations with colleagues in education, in the K–12 sector and, in the business world, to improve the preparation of teachers and principals.
I. Model good practices that increase student learning.
J. Take seriously academic advising that helps students have as much flexibility as possible and is linked to career development services of the institution.

There is no textbook for the course. The inquiry activities are designed to provide student-derived understandings of the concepts and scientific procedures. Students may refer to texts, but soon recognize that they can arrive at a better understanding through their own inquiry activities.

The course utilizes a learning cycle and the success of each student depends on each student sharing the responsibility for learning, the success of the student's laboratory group and at times the whole class, with the professor taking the role of one whose obligation is to skillfully orchestrate the whole process.

Brooks and Brooks[7] have provided a set of conditions that should be present in any setting in which inquiry learning is taking place. This can serve as a metric when monitoring the course for preservice elementary teachers.

Inquiry-Centered Instruction

Inquiry-centered instruction may be described in terms of a set of characteristics shared by teachers adopting this approach.[7]

Such teachers:
- encourage and accept student autonomy and initiative;
- use raw data and primary sources, along with manipulative, interactive and physical materials;
- when framing tasks, use cognitive terminology such as classify, analyze, predict and create;
- allow student responses to drive lessons, shift instructional strategies and alter content;
- familiarize themselves with students' understandings of concepts before sharing their own understandings of those concepts;

- encourage students to engage in dialogue, both with the teacher and with one another;
- encourage student inquiry by posing thoughtful, open-ended questions and asking students to question each other;
- seek elaboration of students' initial responses;
- engage students in experiences that pose contradictions to their initial hypotheses and then encourage discussion;
- allow time after posing questions; and
- provide time for students to construct relationships and create metaphors; and nurture students' natural curiosity.

The Role of Technology in the Course

The laboratory for this course was the first one in the department to use laboratory interfacing as a regular part of a course. Macintosh computers with laboratory interfacing equipment are at each of ten laboratory tables. The Microcomputer-Based Laboratories (MBL) utilize motion detectors to study position and velocity and to build their graphing skills. They use the curve fitting programs to study functions that describe motion: linear, quadratic and trigonomic. They compare the computer-generated results to their own personal calculations and determination of linear equations. One of the goals of the course is to integrate the use of mathematics with the science, so they are perceived as common elements of the language needed to describe the experiments and the physics in a general manner.

The four elements emphasized in the course are experiments, stories, graphs and equations. Students learn to work among these representations of the science and recognize that different students feel at home with different representations, but they, as teachers, need to feel at home with each and skillfully work across these elements. The MBL activities provide high-quality graphs, quick turnaround for experiments that need to be re-run and the ability to analyze the graphs to support the students' verbal or written explanations for the physics involved.

In their study of heat and the conservation of energy, temperature probes are used to study heating, cooling, freezing and melting processes and to recognize some of the properties of ice and water that illustrate so nicely the conservation of energy. Students are asked to design an experiment to determine "how much ice water can melt." After obtaining approval of their plan, students can place a temperature probe in water to which ice is being added. Students notice that the graph of the temperature of the mixture first falls rather swiftly, then begins leveling off at a temperature near $0^{\circ}C$ and that at that point the ice has stopped melting. Careful discussion within the laboratory group leads the students to recognize that a finite amount of water can melt only a finite amount of ice. Laboratory groups create a wide variety of experiments some of which cannot provide an answer to the question. Only when students present their results to the rest of the class do they discover the flaws in their designs, pointed out by the other students, not the professor. One common approach is to simply put a finite amount of ice in water and determine that it has melted. This begs the question of "how much."

The ability to design an experiment, carry out the experiment, make measurements and observations, and then present conclusions based on all of these aspects is a skill that can contribute to the students developing personal confidence that they can do science. This is critical to a willingness to let their own students encounter science in the same way and honors the recommendations of the National Science Education Standards (NSES).[8]

The Context for the Course within a Large Research Department

A large research department has the luxury of offering a variety of introductory courses to different sets of students. This course arose from a one-semester lecture/lab course. A single section was allowed to move entirely into the laboratory, meet two hours at a time, three days a week and have no formal lectures. It was eventually granted its own course identity, was initially taught by John Layman and then was broadened to two sections. Sarah Eno, an assistant professor within the High Energy group who had spent a semester interning with John Layman, taught the second section. She stepped into this course with the blessing of her research group. It has also been taught by a second High Energy assistant professor, a senior member of the faculty and a post-doctoral student.

There is little in the preparation of the average physics professor empowering them to teach in the manner required in this course. It is student centered, activity centered and spends much time guaranteeing student understanding of the limited number of concepts dealt with. There must be recognition that telling what we know in a manner that we think is "clear" does not enhance the understanding of many of the students. Because of what they bring to class and the modeling of successful learning that they experience in class, students can exit with deep understanding of a limited number of concepts, and some of the context within which the learning took place. If students have successfully designed experiments, carried them out and reached conclusions due to their personal skills they will begin to understand the nature of science and how to learn it.

The Conditions for the Course Creation and Transformation

At Maryland I began teaching PHYS 117, the one-semester introductory course, for those seeking only one semester, and used it as the stepping stone for the evolution of PHYS 115, the course for preservice elementary students. We had become part of the Maryland Collaborative for Teacher Preparation, the NSF program designed to improve the preparation of K–8 teachers throughout the state. PHYS 117 served as one of the model sites where inquiry approaches were instituted. The physics department gained credit for supporting this transformation, and the department continues to support two sections per semester, serving over 100 students per year. Those who have now taught the course are two assistant professors, two full professors, one associate professor and one post-doctoral faculty member.

Relationship to the College of Education

The reputation of the course and the skills of the students are monitored when students take their science methods courses. Both full professors are familiar with the physics course. One has brought elementary students and their teachers to the PHYS 115 laboratory to work with the motion detectors so that the present students can see the value of MBL work even for elementary students. The second professor has carried out the formal research on the MCTP program and has funding to continue the research beyond the close of the grant.

These professors find that students arriving from the inquiry-centered physics course are clearly differentiable from other students. They understand inquiry, assume responsibility for their own learning, do not expect to be told everything, and are willing to carry out more open ended and thought provoking activities. They eagerly describe the activities and approaches that enabled them to understand physics. These skills and attitudes coincide with the expectations of the NSES. More will be said about this in the research section of this chapter.

When sections of the course were modified by one senior faculty member bringing them back to a more classical approach, the science-education professors could detect this change when they found the students less excited about their work, and not able to relate their introductory science courses to the skills and understandings expected by the NSES.

Faculty Preparation, Three Approaches

The most successful approach to faculty preparation for teaching in an inquiry fashion was the internship that Sarah Eno participated in. This is best described by her own essay provided in the Appendix.

A second approach was to have one of Sarah's colleagues down the hall brought in. He paid one or two visits to class and then began teaching. He consulted regularly with Sarah and John, and was able to learn the inquiry- and student-centered approach within two semesters.

The third approach was to have a full professor, who had originated the laboratory program for the PHYS 117 class, teach the course. It was he who moved the course more toward a classical approach with more providing information for the students and a somewhat reduced inquiry approach.

The syllabus and laboratory guide are structured to support an inquiry approach, with concern for the merging of the science and mathematics. The character of the software associated with the computer activities also compliments students' development of an understanding of the relationship between graphs, equations and the capacity of an instrument to acquire the data. All of this in support of the student constructing a deep understanding of the concepts and processes of science.

Gaining Departmental Support

In an ideal world, physics departments would recognize their role in providing introductory courses for preservice elementary teachers. If this cannot be done in a separate course, then one of the introductory courses serving this population should be modified. Many of these changes will be of benefit to all students in the course. If there is collaboration between those teaching and monitoring this course and their colleagues in science education, courses at both points in the students' program will be viewed as complimentary and all part of a university-wide teacher-preparation effort. Students should find common expectations across courses and programs.

Sustaining the Teaching/Learning Conditions over Faculty Changes

A special effort must be made within a department to preserve the special character of such a course. It must not be viewed as watered down physics, but as a course that deals with a much broader set of learning/teaching skills and for this reason deals with fewer concepts and deals with these concepts under fundamentally different teaching/learning conditions. Its laboratory-centered approach must be viewed as a special contribution of the department even though faculty could deal with more students at a time in another setting.

Examples of Changes that Can Occur that May Change the Character of the Course

Normally students are asked to read the syllabus and on the first day of class jump right into activities such as using the motion detector without preliminary explanations or detailed instructions. As student understanding develops within the laboratory groups, class discussion refines these understandings. If one chooses electricity as the first activity, students would immediately be asked to use one battery, one lamp and one wire to make the lamp light.

Modifications can occur with a change in teachers. Present to the students a more formal description of What is Inquiry, Teacher Characteristics and Goals for the Laboratory. Adding an Introductory section prior to any physics conceptual work, discussing the Cosmic Voyage, Dealing with Big and Small Numbers, The Metric System, Metric Prefixes, Making Measurements, Conversion of Units and The Greek Alphabet are all helpful. Following this introductory material the first set of physics concepts deal with Electricity and Magnetism. But instead of giving the students a wire, a battery and a lamp, they are asked to first read about How We Know about Electrons. The assumption is that students need more direction instead of allowing students to make conjectures, try things, describe them in their own terms and slowly make a natural transformation to the way we physicists would describe things.

Preserving the character of an inquiry course may be a major challenge within a major department. The course must be viewed as belonging to the department, not the professor assigned to teach it. We are in the early stages of working on this at Maryland.

Research Results, Accounting for Course Influence and Value

Randy McGinnis of the Science Teaching Center at the University of Maryland carried out the formal research of the Maryland Collaborative for Teacher Preparation. He taught some of the sections of the science-methods courses for the preservice elementary teachers and made comparisons of the skills and views of MCTP vs. non-MCTP students enrolled in the methods block.

The research work that I will utilize in describing the relationship that should exist between introductory science and math courses taught in ways that model good inquiry instruction and the science-methods courses preservice elementary students take in the last stages of their programs will be that of Randy McGinnis and Amy Roth-McDuffie. The study, An Action Research Perspective of Making Connections Between Science and Mathematics in a Science Methods Course, focused on six teacher candidates participating in a National Science Foundation-funded undergraduate teacher-preparation program designed to produce specialist mathematics and science upper-elementary/middle-level teachers and on three elementary-education majors with concentrations in mathematics or science. Discussion focuses on the researchers' reflections as prompted by a comparison of the performance of the special teacher candidates and the other teacher candidate participants.

As a result of the teacher candidates' participation in the MCTP reform-based science and mathematics courses, the following research questions were investigated:

1. Are the MCTP teacher candidates distinguished from the non-MCTP teacher candidates in the science content knowledge they bring to their science methods class?
2. Are the MCTP teacher candidates distinguished from the non-MCTP teacher candidates in the beliefs and perceptions they bring to their science methods class concerning:
   a. preparedness to teach science content to elementary students?
   b. an appropriate science learning environment for elementary students?

      c. the structure of mathematics and science?
      d. the rationale for and intent to make connections between science and mathematics in elementary teaching?
      e. the role of science methods in their teacher preparation program?
3. Are the MCTP teacher candidates distinguished from the non-MCTP teacher candidates in the beliefs and perceptions upon completion of the science methods course concerning:
      a. an appropriate science learning environment for elementary students?
      b. the extent to which their science methods professor modeled good teaching of science?
      c. the extent to which they observed their science methods professor making connections to mathematics in his teaching?
      d. the rationale for and intent to make connections between science and mathematics in elementary teaching?

For our work here in Nebraska, I shall report some of the results of just research questions dealing with content knowledge and beliefs and perceptions brought to the methods class that arose from their inquiry-oriented introductory science and mathematics courses.

For research question one, are the MCTP teacher candidates distinguished from the non-MCTP teacher candidates in the science-content knowledge they bring to their science-methods class? We find that the MCTP students had less confidence in their science than the non-MCTP students who had classical courses in science but the MCTP students had higher scores on the science diagnostic instrument. Unfortunately there were no differences in the physical-science portion of the test. This may indicate that the MCTP students retained the skepticism associated with inquiry in making their preliminary judgments.

To answer our second research question ("Are the MCTP teacher candidates distinguished from the non-MCTP teacher candidates in the beliefs and perceptions they bring to their science methods course concerning [a spectrum of areas]"), we analyzed the data we collected from the beginning of the semester teacher-candidate interview. What follows are assertions we generated from a careful reading and comparison of all the participants' responses to the interview questions. These assertions are presented in the order of the sub-sections of the second research question. Included in each are exemplar comments from the participants that support the claims made by our assertions.

a. Content preparedness to teach elementary students. **The MCTP teacher candidates were distinguished from the other teacher candidates by expressing that preparedness to teach young students science content required their being taught content in a manner that modeled good practices. However, as a result of being taught science content by MCTP faculty in a constructivist** manner, which they recognized required a high level of comfort with science content, the MCTP teacher candidates tended to express they felt less prepared as compared with the non-MCTP teacher candidates who were taught content in a lecture-based manner. The non-MCTP teacher candidates expressed a somewhat naive confidence of their content preparedness.

MCTP Teacher Candidate Beliefs and Perceptions. **The distinguishable feature of the MCTP teacher candidates' comments on content preparedness was that they believed their MCTP professors taught content in a manner that modeled good pedagogy, and they could emulate this approach with young learners. They believed this approach promoted lifelong retention of content.**

> Jennifer:
> I think, especially the MCTP classes, we have seen the type of instruction and we have gotten to experience firsthand the way that we want to teach math and science, so that it is not the boring memorization, you know, do that problem ten times, or just memorize the biology and whatever. And I think that we have had a stronger base of the content because it has been taught that way; I think I have learned it more.... I mean, it was more of a displaying of that type of teaching method. Real methods were not taught, you know, about how to teach the subject, but I think more of a display of that type of teaching. (Interview, September)

> Aubrey:
> I think absolutely, totally my Physics 117 was incredible. I think to this day I still have a pretty good knowledge base of what happened in that class and can explain things with some, you know, some level of knowledge and confidence. But I just finished [non-MCTP] chemistry this summer, two sessions, and I probably could not pass any of the exams if they were given to me right now, and that was only about a month ago. (Interview, September)

Non- MCTP Teacher Candidate Beliefs and Perceptions. **A distinguishable feature of the non-MCTP teacher candidates' comments on content was a perception that while they believed that they had gained a sufficient body of science-content knowledge, it had been learned in isolation from a good model of how to teach young students.**

> Patty:
> Science, I would say I am pretty prepared for the elementary level, yes. Middle school, the courses I took are enough—enough I think to probably prepare for middle school. I do not know how much I have retained to be

able to just go in there right now. I mean, I would definitely have to review. (Interview, September)

Kevin:
I had always felt that we had gone through and learned the science content, but that I was never taught how to teach until I got into these classes [method block]. Now I feel quite assured that I will know strategies and ways to deal with teaching that I had felt was really not touched on at all in previous content courses. (Interview, September)

b. A vision of an appropriate science-learning environment for elementary students. **The MCTP teacher candidates expressed a vision of an elementary science-learning environment in alignment with the reform movement (student-centered and problem-based, with an emphasis on students' prior knowledge) that they believed was modeled by their MCTP science content professors. They also could contrast this reform-based vision with a traditional lecture and textbook-based science-content environment. The non-MCTP teacher candidates expressed dissatisfaction with a traditional learning environment based on teacher lecture, but could not express an alternative vision of good teaching for elementary science students except for the increased use of labs involving equipment and manipulatives. Moreover, when they referred to using equipment and manipulatives, the non-MCTP teacher candidates did not indicate that they had developed a vision for how they would use these things or for what purpose.**

MCTP Teacher Candidate Beliefs and Perceptions. **Drawing on their recent undergraduate experience learning science content in MCTP classes, the MCTP teacher candidates expressed a well-developed vision of an elementary science-learning environment that included inquiry, cooperative learning, a concern for students' prior knowledge, the teacher as a facilitator and a commitment to achieving equity between males and females. Furthermore, they indicated they had developed personal theories/rationales for why these modes of learning are appropriate for young learners.**

Karen:
I guess I kind of imagine a classroom setting with the students in groups of four of five; lots of manipulatives at least in the beginning part of the lesson, like an introduction to geometry with the cubes or something like that. And what I have learned, and am finding more and more important, is the discussion taking part in mathematics and science. That it helps the kids understand the concepts more clearly, and it also gives the teacher a chance to assess that way rather than as a quiz with multiplication tables and that

kind of stuff. You can hear what they are talking about and see what kind of level they are at, so I definitely would like to emphasize discussion. "How did you get that answer?" Or if two people got the same answer but they did it differently, "Show how you did it," you know, more like a process than just having the right answer. (Interview, September)

Stephanie:
My first class [in the MCTP] was hands-on with Dr. Layman [introductory physics]. That format is so different, but I feel like that class kind of prepared me for how I want to teach. (Interview, September)

Jessica:
I think learning content has to be non-threatening. I think the group work is good with a lot of hands-on materials. I think it should be something that it seems like it is a situation that is fair to both males and females.... My vision of my ideal science classroom, I would have lots of living things all around the class—animals, fish, plants, just all kinds of stuff all over the walls. I would have all kinds of different areas that students can move to and explore and learn things, books that they can look at, things that they can look at, things that they are interested in, lab tables, lots of equipment—a student-centered, really nice environment where they would be learning by doing things hands-on. Group work—manipulatives, experimenting, finding things out on their own. (Interview, September)

Non-MCTP Teacher Candidate Beliefs and Perceptions. **In the context of their recent undergraduate experiences of learning content in a lecture-based manner that they believed was inappropriate for young learners, the non-MCTP teacher candidates' alternative vision of good pedagogy for young learners was one based on instances of good teaching in their own K–12 educational histories or on brief field-based education experiences observing young students. These alternative visions were not thoroughly developed.**

Anna:
As an elementary student, I always liked the practical experiments. Like, when I was in second through fourth grade I did not speak much English, and with the experiments and laboratory work, I would learn through observing the lab, the experiment, the actual experiment. I could not read or understand, so I only learned through observation. (Interview, September)

Patty:
I think in the elementary level, I think manipulatives are real effective. [pause] I have done a lot of one-on-one with kids. Most of my experience

[as a parent volunteer in an elementary school] has been with second grade, and I have done a lot of one-on-one or working in small groups, and it seems like it is much easier to show them using something than to just try and tell them, so definitely manipulatives is an effective way. (Interview, September)

c. The structure of mathematics and science. **The MCTP teacher candidates brought to their science methods a shared vision of the structure of the disciplines of the science that was characterized by being in alignment with current philosophical thought on the structure of the disciplines. The non-MCTP teacher candidates expressed a limited vision of the structure of science and mathematics that in many ways conflicted with current philosophical thoughts on the disciplines.**

MCTP Teacher Candidate Beliefs and Perceptions. **The MCTP teacher candidates expressed a shared perception that mathematics and science were similar structurally. The similarities included: intellectual pursuits based on curiosity, ways to better understand logical systems and the physical universe, with the primary aim to improve the quality of life through solving problems. They perceived the disciplines as different with science more limited by a tentativeness nature of knowledge and with mathematics more structured and static that led to more conclusive answers.**

Bob:
They are both problem solvers. Both of them are used in solving problems, in trying to improve the quality of life, or to understand our world. I think...I...I seem to think math is more structured than science, just more rules governing math than science. (Interview, September)

Karen:
Both mathematics and science solve mysteries. And with math it seems like there is always an answer, sometimes in science there is not. There might be, like, a theory. It seems like math doesn't change that much. (Interview, September)

Aubrey:
I think they both tend to be inquiry based. I think there is a lot that can be done with that as far as building on the student's prior knowledge and just working into their questions and their desires for what they want to learn. I think that the focus can be more on the terms and the ways of using, you know, the knowledge. I think they are both very similar in how you can use them to discover things and hands-on activities. (Interview, September)

Jennifer:
I think in both math and science, there are a lot of things that we don't know... science seems to almost not work without having math as part of it.... I would think that, also, a true scientist would have to have some mathematical background to be able to do some of the experiments and that is how I see a true scientist as an experimenter.... I think that a true mathematician might be able to, you know, work in his profession without a whole lot of science background. (Interview, September)

Non-MCTP Teacher Candidate Beliefs and Perceptions. **The non-MCTP teacher candidates indicated they held the perception that mathematics was distinguished from science by mathematics being a static discipline concerned with finding conclusive answers to algorithm-based problems. Science was perceived as a growing area of knowledge based on inquiry. There were some claims of similarity of the disciplines as both being based on formal reasoning skills.**

Anna:
What similarities do I see? They both have some research. Both science and math you usually hypothesize. Math is straightforward. It is a one-answer, one-solution problem. (Interview, September)

Patty:
I am thinking that science is much broader [than mathematics] because it is always changing. I know there has been a lot of new math that has come up over the last ten or 15 years, but the basis of math is one plus one is always two. It is always going to be, and always has been.... I want to say formal reasoning for both. (Interview, September)

Kevin:
They share logic skills. Science can be hands-on, real life and then math is really just a bunch of symbols when it comes down to it. (Interview, September)

d. Rationale for and intent to make connections between science and mathematics in elementary teaching. **The MCTP teacher candidates evidenced considerable reflection based on the firsthand MCTP experience of learning science and mathematics in a connected manner for a rationale making connections between science and mathematics. They intended to make extensive connections between the disciplines in their future practices. The non-MCTP teacher candidates were characterized by not having reflected on a rationale for making connections between the disciplines nor having experienced** learning the disciplines in that manner except in cases where mathematics was used as a tool in science. They expressed a willingness to make connections between mathematics and science but based that connection solely on the use of mathematics as a tool.

MCTP Teacher Candidate Beliefs and Perceptions. **The MCTP teacher candidates brought to their science-methods course the ability to articulate a rationale for making connections between science and mathematics based on extensive prior experience of learning the disciplines in that manner. Through their MCTP experiences, they perceived mathematics and science to be so intrinsically connected that they had difficulty conceiving teaching them as separate subjects. Their rationale included the belief that both disciplines could contribute, and in the case of mathematics, assist the other, in developing a better holistic understanding of an area of interest.** They professed a shared intent to make extensive connections between the two disciplines in their future teaching practices.

> Stephanie:
> Well, I pretty much think that mathematics and science are interconnected. I mean, if you think about the formulas in science, you are learning all that in math, also.
>
> Jessica:
> I think that one of the reasons Stephanie might think that and that I might think that, too, is just because we have been learning it that way, for the past four years (I know I have anyway). And so I say, "Oh yeah, math just fits in with science, and science just fits in with math naturally. How would they not?" And maybe some people do not see that and do not emphasize it. I do not know if it is something that we have to emphasize so much and try and make a point of doing it because we are just so used to doing it anyway, and it is just going to naturally kind of fit in.
>
> Karen:
> Making connections keeps things as a whole, and you know, learning parts, and parts, and parts, and parts that is just a bunch of parts, but if you make connections all across the board, especially with math and science, because they relate so much, it just keeps everything like a nice package all wrapped up.
>
> Jessica:
> I think you can make connections between mathematics and science using calculators, graphing, all sorts of graphs, all kinds of graphs that you could

do for different things in science. Doing different trials, and making graphs of your findings type things. I mean, math naturally comes out in science that way. For math activities, you could give them activities, too. An activity I had in an MCTP math class comes back to my mind. It was about learning about shadow lengths and how we could determine how the people in the past could determine that the earth was round and the distance around the earth by a change in shadows. I mean, that is an example of a way that would relate science and math together, and you could do it in a math class, and kids might not think they were learning science, but they would be learning science just by measuring shadows and that sort of thing.

Aubrey:
I think mathematics and science can be connected largely by not calling it a math lesson or a science lesson. I think dealing with the topics and letting them flow into the different subjects sort of leads to an integration without forcing it. And questioning, open-ended questions, and probing questions that would lead them to kind of make those discoveries in their minds and draw their experiences from both together. I want to set up things so, like, if my units are more interdisciplinary, so then the connections, hopefully, become obvious at least in a way that the kids are going to feel like they can go home and say, "Mom, I did this today. This was math, but you know what? It was also science and it was really fun and important."

Stephanie:
You do not have to say, "Look, there is a interconnection between these two subjects." It is going to come out naturally.

Non-MCTP Teacher Candidate Beliefs and Perceptions. **The non-MCTP teacher candidates brought to the science-methods course a restricted rationale for making connections between mathematics and science. While they voiced a willingness toward attempting to make connections between science and mathematics, they based the connection fundamentally between science and mathematics on mathematics use as a tool in science.**

Patty:
Oh, this one I will have to think about...I am sure I could come up with lots of ways to tie them together, I just cannot think of any right now.

Kevin:
I think it is important to make connections between mathematics and science. . There is quite a large connection between the two of them. You can always figure out science properties by doing the experiment, but then it

is usually the math that is used to prove them.... Hopefully I will learn how to connect mathematics and science this semester [during the methods block].

Anna:
Usually, when you collect data from science, you are actually doing the math, because most of the experiments want you to find the average.

Kelly:
Well, I think it would be easier to show the connections going from science to math for me. To show that how—I cannot think of an example—but when they have done an experiment and they had to, like, say write the results down, and they have made a graph or something and then you can connect that to the math.

e. The role of science methods in their teacher preparation program. **The MCTP teacher candidates brought to the science methods class an inclusive vision of teacher preparation program composed of a seamless linkage between their undergraduate content courses and their science methods course. As a result of being taught content in a manner that modeled good pedagogy, they had a vision of how they wanted to teach. However, they recognized that the science methods course was essential to teach them the skills and knowledge base to enact that vision of teaching. The non-MCTP teacher candidates brought to science methods a vision of content classes taught in a manner they believed was inappropriate for young learners. They saw the science methods as their first opportunity to gain skills in teaching science appropriately.**

MCTP Teacher Candidate Beliefs and Perceptions. **The MCTP teacher candidates held the vision of science methods as performing an important next step role in their teacher preparation program by assisting them in enacting their vision of teaching content to young learners appropriately. They believed the primary purpose of science methods was to give them the opportunity to develop the strategies and knowledge necessary to adapt what they previously observed their professors doing in science content classes to lessons for young learners.**

Stephanie:
I am hoping to actually learn how to tie everything together....We are going to be learning about the different methods of teaching. That is what I am hoping to gain from it.

Jessica:
I was thinking I would learn in science methods how I am going to use what I learned, take it to a classroom and fill up the day teaching what I know. What I will actually have to do to get across the things that I need to get across to the students without having to tell them these things directly.

Karen:
Learn how to do lesson plans.... I am concerned about day-to-day, what do you do? I mean, how far in advance are you prepared? You know, I have this image, that 10-year veteran teachers have their whole year planned out, but how much can I possibly get done in just this semester to even prepare myself for the 12 weeks of teaching should I have? That is kind of one of the things I am hoping to get out of methods is that I will feel ready to go in and student teach.

Bob:
It is the preparation, getting lesson plans together, knowing where you are going to go with it. I am hoping to learn all of that.... I guess in methods I am hoping to learn planning and organization, and how to present the material and all of that lesson plan type thing. That is where we are stuck.

Non-MCTP Teacher Candidate Beliefs and Perceptions. **The non-MCTP teacher candidates saw the science-methods course as their first opportunity in their undergraduate program to focus on the teaching of science to young learners in an effective and appropriate manner. They expressed interest in learning the strategies to teach science as if they were content [independent].**

Kelly:
Oh, what I hope to gain in science methods is knowledge of the strategies to teach. This is the first time that they have come up.

Anna:
How to come up with questions to ask, because if I was just to give a lesson right now, I would not go too deep with the details to ask how would they get that. So I guess so far I have learned I need more to learn.

Kevin:
Just the different strategies, the different ways of looking at certain topics that are associated with difficulties for children to learn certain topics. How to get around them, how to set them up with different features and things like that.

Additional Characteristics

Let me turn to two key characteristics of an inquiry-based introductory course for preservice teachers, journaling and summarizing.

Journaling

Journaling can play a key role in providing students an opportunity to metacognate on their personal progress in learning. It is an opportunity to look at the big picture and summarize their thoughts and feelings about their own progress in the course. Here are examples from PHYS 117.

Dear PHYS 117 listserv members. Almost all of you have signed on to our listserv and I want to welcome each of you to our communication venture.

Journal #1

Our First Journal is: In PHYS 117 we are trying to create a "learning community." If we are successful, what obligations do you have as a student, and what obligations do we have as instructors?

The response is to be one page or less. This can mean about one page of computer screen response or one page of printout had you done this. Be sure that you identify your response in terms of which Journal response yours is (this is Journal #1), the date and your name.

Good luck. Remember you may submit your response to me personally at jl15@umail.umd.edu, but it would be nice to share it with the other members of our learning community as well.

John Layman

PHYS 117A Journal #2

We have almost completed our study of heat and the conservation of energy in which we tried to create an equation or formula for making calculations or predictions of mixing experiments that eventually included mixing of different substances. Some managed to find a rule or equation, while others said that they "did not do equations." Others said that they had no difficulty in using an equation once someone gave it to them.

What is your present view of our work with equations, whether recognized or borrowed, and the role they play in your understanding of heat energy measurement?

PHYS 117A Journal #3

We design and carry out experiments and make skilled observations of their results in our class regularly. What is your view of experiments and do you feel at home with experiments?

Summarizing

The second feature is the mechanism of bringing closure to a particular broad concept. After students have carried out a small series of interrelated laboratory activities they are asked to summarize their observations and understandings. Example from motion activities with a Fan Cart in PHYS 117.

Activities with a Fan Cart.
M3.1: Visual Observations and Telling a Story
M3.2: Monitoring the Fan Cart with a Motion Detector
M3.3: Learning More from Velocity Time Graphs
M3.4: Curve Fit II

M3: Summary

You have now observed, told a story about, recorded, equationed and curve-fitted a new kind of motion. What are the special features of this motion?

What would you need to do if you were to walk in front of the motion detector to produce the same kind of graph?

Closing Comment

The expectations of the NSES speak to all elements of our community. Colleges and universities have an obligation to prepare teachers able to meet the standards, and departments have their role to play. The physics community is one of the most successful in terms of studying teaching and learning. We should translate this into a major contribution to the preparation of the next generations of elementary teachers, all of whom will be colleagues in the teaching and learning of science.

References

1. *Powerful Ideas In Physical Science*, American Association of Physics Teachers, One Physics Ellipse, College Park, MD 20740, 1996, NSF DUE 9496330.
2. Maryland Collaborative for Teacher Preparation, NSF DUE 9255745.
3. Title: NSF 96-139—SHAPING THE FUTURE: New Expectations for Undergraduate Education in Science, Mathematics, Engineering, and Technology, Type: Report, NSF Org: EHR/DUE, Oct. 3, 1996, File: nsf96139.
4. Ibid, p. 8.
5. Ibid, p. 8.
6. Ibid, p. 73.
7. Brooks, J.G., and M.G. Brooks, In Search of Understanding: The Case for Constructivist Classrooms, Alexandria, VA, Association for Supervision and Curriculum Development, 1993.
8. National Science Education Standards, National Research Council, National Academy of Sciences, 1995.

The Role of Physics Departments in Preservice Teacher Preparation: Obstacles and Opportunities*

Jose Mestre
University of Massachusetts

Introduction

What role should university physics departments play in the education of prospective K–12 science teachers? This is an important question to answer in view of various national agendas to improve the quality of science instruction for K–12 students. Although it would be easy to argue that physics departments have played an important role in offering content courses to prospective teachers, it could also be argued that the pedagogy used in teaching such courses has done little to encourage the type of teaching consistent with research on learning that we would like prospective teachers to adopt. In this article I begin by describing the attributes of an ideal, generic[1] physics course for prospective K–12 teachers. Next I will attempt to justify why the particular attributes of this course were selected based on learning research. I then provide some specific examples of successful attempts to incorporate learning research in the design of university physics courses. A discussion follows of why, at the present time, the vast majority of physics departments are not poised to offer courses containing many of the attributes described, and suggestions are provided for how this situation could be alleviated. I conclude with a discussion of the obstacles and opportunities for reforming the role of physics departments in the education of prospective K–12 teachers.

Attributes of "The Ideal" Physics Course for Prospective Science Teachers

The list of attributes that I will provide is not intended to be complete, and is very likely somewhat idiosyncratic; someone else's list will likely differ, but if any two lists that are supposedly justifiable based on cognitive research findings are compared, there should be considerable overlap. Further, the list is intentionally general and will not differentiate between courses aimed at the elementary-, middle-, or high-school levels. Finally, there is also no hierarchy intended to the list of attributes.

- The course should integrate physics content and pedagogy. When pedagogy and content are taught separately, they are seldom integrated. An ideal course for prospective teachers would integrate the content with effective ways of teaching that specific content, with the goal being to develop what is termed "pedagogical content knowledge" for teaching the specific subject (discussed later in more detail).
- The classroom environment should encourage students to construct and make sense of their physics and pedagogical knowledge. Although teachers can facilitate learning, students must do the learning themselves. Evidence from research on learning suggests that "teaching by telling" is not the most efficient way to help students learn science. Students must learn science content in ways that make sense to them and their understanding of that science must be consistent with scientists' current models for how the physical and biological world works. Classroom environments where students are actively engaged (e.g., inquiry learning, cooperative group learning, hands-on activities) are helpful in achieving this goal. In this type of classroom environment, the teacher plays the role of learning coach, rather than authority figure dispensing knowledge. The same holds for the pedagogical knowledge needed by the prospective teachers to teach the science content effectively; students must learn effective pedagogical practices by actively practicing them with peers in the class, and as stated in the previous attribute, by integrating pedagogical knowledge with their science content knowledge.
- An ideal course should be content-rich. Clearly any physics course for prospective teachers has to be based on physics content, but at the same time it should not be so laden with content that it becomes a race to survey as many topics as possible. The emphasis should be on understanding in depth a few major topics rather than the memorization of lots of facts about many topics; the former has lasting value, the latter is quickly forgotten after the course is over.
- Ample opportunities should be available for learning "the process of doing science." Doing science not only requires lots of content knowledge but also knowledge about the processes involved in scientific investigation, that is, knowledge of the process of science. Students should therefore use apparatus, objects, equipment and technology to design experiments and test hypotheses. Rather than "cookbook" labs where students are guided at every stage, students should develop the ability to pose answerable questions and to use lab equipment to design experiments to answer their questions. There is a fine line between leaving students totally alone risking frustration and floundering, and providing just enough guidance so that they make suitable progress. Here we are striving for the latter, with the goal being to illustrate the building of explanatory models based on the physics being learned.
- Ample opportunities should be provided for students to apply their knowledge flexibly across multiple contexts. Physics is perhaps the only science in which a handful of concepts can be applied to solve problems across a wide range

of contexts. Unfortunately, the research literature suggests that when people acquire knowledge in one context they can seldom apply this knowledge to related contexts, which look superficially different from the original context, but which are related by the major idea that could be applied to solve or analyze them. The implications are that students should learn to apply major concepts in multiple contexts in order to make the knowledge "fluid."

• Students should be assisted in organizing content knowledge according to some hierarchy. For students to be able to learn lots of things about a topic, to recall that knowledge efficiently and to apply it flexibly across different contexts requires a highly organized mental framework. A hierarchical organization, where the major principles and concepts are near the top of the hierarchy and where ancillary ideas, facts and formulas occupy the lower levels of the hierarchy but are linked to related knowledge within the hierarchy, is one possessed by those achieving a high level of proficiency in a field.

• Students should practice constructing qualitative arguments to explain phenomena and/or experimental findings and to highlight the major components of problem solutions. A lot of the knowledge that scientists possess is referred to as "tacit knowledge," or knowledge that is used often but seldom made explicit or verbalized (e.g., when applying conservation of mechanical energy, one must make sure that there are no non-conservative forces doing work on the system). Tacit knowledge needs to be made explicit before students can recognize it, learn it and apply it. One way of making tacit knowledge explicit is by constructing qualitative arguments using the physics being learned. By both constructing qualitative arguments, and by evaluating others' arguments, students can begin to appreciate the role of conceptual knowledge in "doing science." This type of activity also helps develop other worthwhile skills, such as ability to communicate both orally and in writing, ability to organize and prioritize knowledge, ability to develop and use the vocabulary of the discipline and, perhaps most important, the ability to use and refine the type of reasoning that is valued in physics.

• Attempts to teach students metacognitive strategies should be integrated throughout the course. Metacognition is a term used in the cognitive literature to refer to people's abilities to predict not only their ability to perform tasks but also their current levels of mastery and understanding. In essence, metacognition refers to thinking about the learning process. By helping students to be self-reflective about their own learning, they can learn how to learn more efficiently. For example, when stuck trying to solve a problem, asking oneself questions such as "What am I missing or what do I need to know to make progress here?," "Am I stuck because of a lack of knowledge or because of an inability to identify or implement some procedure for applying a principle/concept?," are often helpful in deciding on a course of action. After solving a problem, reflecting on the solution by asking questions such as "What did I learn that was new by solving this problem?," "What were the major ideas that were applied and what is their order of impor-

tance?," "Am I able to pose a problem in an entirely different context that can be solved with the same approach?," help one monitor mastery and understanding of the topics being learned.
• Formative assessment should be used frequently to monitor students' understanding and to help tailor instruction to meet students' needs. Formative assessment, or assessment for purposes of providing feedback to both students and instructors to help guide teaching and learning, provides valuable information to both students and instructors; formative assessment helps students realize what they do not understand and helps teachers to craft tailored instructional strategies to help students achieve the appropriate understanding. Evidence indicates that without frequent probing for student understanding, students often construct inaccurate or incomplete understandings that interfere with subsequent learning. This practice would also model for prospective teachers a very powerful pedagogical strategy that they should adopt when they become teachers.

Why These Attributes?

Over the last two decades, cognitive research has made great strides in helping us understand the learning process. It should not be surprising that findings from research on learning are very suggestive about the ingredients that should be present in effective instruction. Several review articles have done a good job discussing the implications of this research to the teaching of physics.[2-10] Perhaps the best synthesis of research on learning is contained in a recent report from the National Research Council titled How People Learn: Brain, Mind, Experience and School,[11] which goes beyond synthesis and provides examples of how learning research can be applied in teaching. In this section I will discuss the attributes above in view of findings from research on learning.

The Nature of Expertise

Much of what is known about knowledge acquisition, storage in memory and application to solving problems has come from studies of experts engaged in problem-solving tasks in their domain of expertise. Experts have extensive knowledge that is highly organized and used efficiently in solving problems, and so cognitive scientists have focused on characterizing the organization, acquisition, retrieval and application of experts' knowledge.[11] Among the salient findings is that experts' knowledge is highly organized.[8,12-14] The organization is hierarchical, with the top of the hierarchy containing the major principles/concepts of the domain; ancillary concepts, related facts, equations, etc., occupy the middle to lower levels of the knowledge pyramid. Because of the highly organized nature of their knowledge, experts are able to access their knowledge quickly and efficiently. Further, procedures for applying the major principles and concepts are closely linked to the principles and retrieved with relatively little cognitive effort when a major principle is accessed in memory. This allows experts to focus their cognitive efforts on analyzing and solving problems, rather than on searching for the appropriate "tools" in memory needed to solve the problems. Knowing more, by virtue of having an efficient organizational structure of the knowledge, means that it requires relatively little effort for the expert to learn even more about their area of expertise since new knowledge is integrated into the knowledge structure with the appropriate links to make recall and retrieval relatively easy.

Two of the attributes for an ideal physics course above, that the courses should be content rich and that the organization of the students' knowledge is as important as the knowledge itself, are suggested by the research findings just reviewed. Learning a few topics in depth, especially if effort is devoted to organizing the knowledge, has more lasting value than covering lots of material with little attention toward how it should be organized in memory. Students are often adept at memorizing information for use in tests, but short-term memory is not what we have in mind for the physics knowledge needed by prospective teachers (or physics majors for that matter).

Experts also approach problem solving differently from novices.[15] For example, when asked to categorize problems (without solving them) according to similarity of solution, experts categorize according to the major principles that can be applied to solve the problems (e.g., conservation of momentum), whereas novices categorize according to the superficial attributes of the problems (e.g., according to the objects that appear on the problem statement, such as "pulleys" and "inclined planes"). When asked to state an approach they would use to solve specific problems, experts discuss the major principle they would apply, the justification for why the principle can be applied to the problem and a procedure for applying the principle. In contrast, novices jump immediately to the quantitative aspects of the solution, discussing the equations they would apply to generate an answer.

This research suggests that the tacit knowledge that experts use to solve problems should be made explicit during instruction, and that students should actually practice applying this tacit knowledge while solving problems. If one believes that learners learn by constructing knowledge (see next section), however, this cannot be accomplished by simply telling students how major ideas apply to problems—students need to engage actively in applying and thinking about how the big ideas are relevant for solving problems so that they become internalized as useful problem-solving tools. Several research studies suggest that it is possible to get introductory physics students to attend to "high-level" knowledge (as opposed to simply manipulating equations). For example, studies indicate that students are more likely to focus on conceptual knowledge in problem categorization and problem solving tasks following "treatments" in which they spent time analyzing problems qualitatively before attempting a quantitative solution.[16–21]

Current View of Learning

The contemporary view of learning is that individuals actively construct the knowledge they possess. Constructing knowledge is a life-long, effortful process requiring significant mental engagement from the learner. One of my favorite analogies that brings home the point that learning has to be done by an individual, and can at best be facilitated by teaching is the following quote from an article by Orville Chapman in a special issue on The Science of Learning in the Journal of Applied Developmental Psychology:[22]

Learning science is like losing weight. You can hire a coach to instruct you, you can read books about it, and you can get lots of free advice on how to do it, but in the end you must do it.

In contrast to the "absorbing knowledge in ready-to-use form from a teacher or textbook" view of learning, the "constructing knowledge" view has two important implications for teaching. One is that the knowledge that individuals already possess affects their ability to learn new knowledge. When new knowledge conflicts with resident knowledge, the new knowledge will not make sense to the learner and is often constructed (or accommodated) in ways that are not optimal for long-term recall or for application in problem-solving contexts.[23–27] For example, when children who believe the Earth is flat are told that it is round, they accommodate this to mean that it is round like a pancake, with people standing on top of the pancake.[28] When subsequently told that the Earth is not round like a pancake, but rather round like a ball, children envision a ball with a pancake on top, upon which people could stand (after all, students reason, people would fall off if standing on the side of a ball!). Thus, prior knowledge and sense-making are prominent in the constructivist view of learning.

The second implication is that instructional strategies that facilitate the construction of knowledge should be favored over those that do not. Sometimes this statement is interpreted to mean that we should abandon all lecturing and adopt instructional strategies where students are actively engaged in their learning. Although the latter goal is certainly desirable, the former is an overreaction. It is certainly true that, under the right conditions, lecturing could be a very effective method for helping students learn, but wholesale lecturing where students are passively writing down notes is not an effective means of getting the majority of students engaged in constructing knowledge during class time. Hence, instructional approaches where students are discussing physics, doing physics, teaching each other physics and offering problem-solution strategies for evaluation by peers will facilitate the construction of physics knowledge.

The Relationship between Content Expertise and Teaching

Although being an expert in a discipline is a necessary condition for teaching that discipline effectively, it is not a sufficient condition. An effective instructor also has a wealth of "pedagogical content knowledge," which includes knowledge about the types of difficulties that students experience, typical paths that students must traverse to achieve understanding and potential strategies for helping students overcome learning obstacles, all of which are discipline-dependent.[11,29,30] Pedagogical content knowledge also differs from knowledge about general teaching methods, which are often taught within "methods" courses outside of the science discipline. What cannot be learned in isolated methods courses about teaching a specific discipline such as physics are such things as the types of assignments that are best suited for teaching particular topics, the types of assessments that are best suited to gauge students' progress and to guide instruction, and the way to structure classroom discussions to highlight and clarify new ideas, as well as integrate them within the students' knowledge structure. In short, there is an interaction between knowledge of the discipline and the pedagogy for teaching that discipline, which for the experienced instructor results in a "cognitive road map" that guides the instructor while teaching. All of this raises the question of how prospective physics university professors are supposed to develop pedagogical content knowledge when nearly all of the emphasis in PhD graduate training is on content; this is an issue to be discussed in a later section.

Assessment in the Service of Learning

Most assessment carried out in university science courses is "summative," or intended to sum-up the competence of the students and assign grades. Largely missing from science classrooms, especially large lecture courses, is formative assessment intended to provide feedback to both students and instructors so that students have an opportunity to revise and improve the quality of their thinking, and instructors can tailor instruction appropriately. Perhaps the biggest deterrent to using formative assessments in science classes is that instructors lack techniques for using continuous formative assessment in ways that are unobtrusive and seamless with instruction. The age-old techniques of asking a question to the class and asking for a show of hands has been tried by most but does not work well since few students participate in the hand-raising largely due to lack of anonymity. Because research on learning indicates that all new learning depends on the learner's prior learning and current state of understanding, to ignore students' current level of understanding during the course of instruction is perilous.

In small classes it is not difficult to shape teaching so that two-way communication takes place between the instructor and the student. For example, one very effective method of teaching physics to small classes perfected by Minstrell[31] involves the class-wide discussion led by the teacher. Here students offer their reasoning for evaluation by the class and by the instructor, with the class format taking somewhat the form of a debate among students, with the instructor moderating the discussion and leading it in certain directions with carefully crafted questions. In large enrollment classes the advent of classroom communication systems has allowed the incorporation of a workshop atmosphere, with students working collaboratively on conceptual or quantitative problems, entering answers electronically via calculators and seeing the entire class' response in histogram form for discussion.[32–34] With this approach, the histogram serves as a springboard for a class-wide discussion where students volunteer the reasoning leading to particular answers, and the rest of the class evaluates the arguments. The instructor moderates, making sure that the discussion leads to appropriate understanding. Other approaches, such as Law's Workshop Physics[35] and McDermott's Physics by Inquiry,[36] are intended for small classes where students engage in hands-on learning, and the instructor circulates making sure students make suitable progress; this allows individualized formative assessment since the instructor can serve as diagnoser/tutor as s/he circulates about the room.

Transferring Knowledge Flexibly across Different Contexts

Transfer, which refers to the ability to apply knowledge learned in one context to a novel context, is difficult to achieve with traditional instruction.[11] In physics, we have all heard complaints from instructors that students do not apply what they learn in math class in physics class. Research suggests that transfer is not easy to accomplish. A classic example of this difficulty is illustrated in Figure 1. Although few of the college students participating in that experiment were able to see the relevance of the first problem to solving the second problem on their own, most could apply the major idea used to solve the first problem to the second problem once told that there was a connection.

In my own work with problem posing,[37] I found that when asked to pose solvable, textbook-like problems from problem scenarios,[38] students were quite constrained due to their inability to find multiple contexts in which to apply concepts. For example, when posing problems that incorporated "conservation of mechanical energy" students always used the same context, namely an object undergoing free fall in the earth's gravitational field. Because of this, it was often impossible for them to match other pieces of the concept scenario with a problem posed within this context [e.g., in the concept scenario in the footnote below, it was impossible for students to pose a problem in which a falling object collides and sticks to another object (thus far the first two parts of the scenario are satisfied, namely conservation of mechanical energy followed by conservation of momentum), and then have potential energy increase and kinetic energy decrease]. The concept scenario below could have been easily satisfied within a context containing a spring, which, besides gravitation, is the other major system studied in introductory physics that conserves mechanical energy.

Research suggests that several features of learning affect transfer.[11] First, the amount of learning clearly affects whether the knowledge is available to be transferred, and this depends on the time on task and the student's interest and motivation to learn the material. The context in which the knowledge is learned is also pivotal in terms of ability to transfer; if knowledge is learned solely in one context, it will be unlikely that it is transferable to other contexts. This implies that as new knowledge is learned, students should be assisted in considering the multiple contexts in which it applies, and in linking the knowledge to previously learned knowledge. Finally, new learning involves transfer from previous learning, and often previous learning can interfere with ability to transfer knowledge appropriately to new contexts (the physics education research literature on "preconceptions" or "alternative conceptions" is an archetypal example of this).

Metacognition: Making Defensive Learners

Research suggests that transfer can be improved, that is, ability to use knowledge in new contexts without the need for explicit prompting, by the use of metacognitive strategies.[39–41] Metacognitive strategies refer to strategies for helping learners become more aware of themselves as learners and include ability to monitor one's understanding through self-regulation, ability to plan, monitor success and correct errors when appropriate, and ability to assess one's readiness for high-level performance in the field one is studying.[11] Reflecting about one's own learning is a major component of metacognition and does not occur naturally in the physics classroom both due to lack of opportunity and because instructors do not emphasize its importance. It is common to hear during a one-on-one session with a student during office hours the comment, "I am stuck on this problem," but when asked about more specificity about this condition of "stuckness," students are at a loss to describe what it is about the problem that has them stuck, and often just repeat that they are just stuck and cannot proceed. If during instruction we were to take the time to suggest why, and how, students should reflect about their learning, there would be fewer incidents of the "stuck" condition, since students would be able to identify what they are missing that would allow them to proceed.

Promoting that students reflect on their learning is also pivotal in physics courses that deviate from the norm in pedagogy. Despite research evidence that students learn best when actively engaged, the norm in college physics instruction is the lecture in which most students are passively taking notes. Courses that attempt to get students to work collaboratively, or that try other techniques to engage them, are often viewed by students as being deviant, and thus to be simply tolerated rather than invested in. In cases such as these, instructors should communicate with students why the course is being taught the way it is, and explain how research on learning suggests that the approach being used is superior to the teach-by-telling approach. Only by getting students to reflect about their learning, and by accruing evidence that in fact the "active learning" approach is making them learn more than a lecture approach, will students begin to buy into the approach and become active participants rather than simply tolerant participants.

Sample Courses Containing Some of the Attributes Listed Earlier

In this section I will describe three courses taught in the Department of Physics at University of Massachusetts that incorporate many of the attributes described earlier. Although two of the three courses were not specifically designed for preservice teachers, they nevertheless apply sound learning principles that should be used in teaching any student.

Active Learning in Large Lectures

Large introductory physics courses, especially service courses for engineering or life-science majors, are notorious for problems such as lack of student interest and engagement, low student achievement and poor attendance (often reaching as low as 30% one month after the beginning of a course). With the advent of classroom communication technologies (CCTs), the UMass Department of Physics has been offering a more pedagogically rich experience to students in large enrollment physics courses, while at the same time promoting an atmosphere of collaboration and collegiality among students, and between students and the professor.[32-34] We now use two types of CCTs: Classtalk and the Personal Response System (PRS). With both technologies, an instructor is able to present questions/problems to the class and collect answers electronically via hand-held devices (either a hard-wired calculator or a wireless infrared transmitter). A computer under the control of the instructor tabulates and displays a histogram of the class' responses and keeps a record of all students' responses during the class period.

Class time is spent very differently within a CCT-taught class. As in any other large class, there is always background noise and activity, especially at the beginning of each class. To get started, the instructor projects a question onto a screen, and almost immediately the whole room becomes very quiet, as students read and interpret the question posed. Soon, the noise level in the room starts to build as groups of students work collaboratively in formulating an answer to the question. The discussion eventually reaches a crescendo with everyone in the class debating the question. When the noise level begins to drop again, the instructor asks students to input their answers. Once all the answers have been collected, a histogram of the class' responses is projected and serves to focus students' attention on the front of the classroom. Based on the distribution of answers, the instructor can decide what to do next. It might be a targeted mini-lecture, a demonstration, or a class discussion of the reasoning behind each answer. The instructor in this scenario is more of a learning coach, facilitating and directing learning where needed.

This instructional approach enhances the communication among students, as well as between the students and the professor. By viewing the resulting histogram of the class' response to a question and by presenting their reasoning for evaluation by the class, not only are students able to monitor their own understanding, but the instructor also receives invaluable feedback that s/he can use to tailor instruction to meet students' needs. Further, by facilitating a shift from a passive, teacher-centered (i.e., lecture) classroom toward an interactive, student-centered classroom, CCTs help to create a classroom environment that accommodates a wider variety of learning styles, making the learning of difficult subjects like science a much more positive experience for students.

CCTs can also provide motivation for implementing active learning in large classes. We have found it difficult without CCTs to engage students in collaborative group work in large lecture courses. Entering their own answers and then viewing the entire class' response motivates students to become engaged in group work. We find that without CCTs, students would rather listen to an instructor lecture, but with CCTs students want to know how they are doing relative to the rest of the class, and often tell us that they are relieved to find out that they are not the only ones getting the wrong answer. Although CCTs can never re-create the atmosphere of a 10 to 20 student class, they are an excellent tool for creating a small-class atmosphere in a large lecture class.

We have found four additional outcomes since the start of CCT-based instruction: 1) Attendance has increased dramatically. Since CCTs give us the ability to hold students accountable for attending class, we find that attendance has gone from 35% to 45% when taught traditionally with lectures, to 80% to 90% when taught with CCTs. We believe that this increased attendance is largely responsible for reducing the fail-rate in large service courses. For example, in a comparison of the same two introductory physics courses for engineering majors over nine offerings, four without CCTs and five with CCTs, all taught by the same instructor, the fail-rate averaged 12% without CCTs vs. 5% with CCTs. 2) There has been a shift in CCT-based active learning classes from presenting content to helping students formulate their own understanding of the content. 3) Students' ability to formulate and present arguments based on what they are learning improves over a semester of CCT-based instruction, as well as students' ability to evaluate critically other students' arguments. 4) Data still under analysis from on-going research suggest that there are only slight improvements in problem-solving skills (as measured by traditional problem-solving tests) with CCT-based instruction compared to traditional lectures but that conceptual understanding increases with CCT-based instruction.

In summary, this instructional approach incorporates several of the attributes described earlier.[42] For example, students' ability to construct knowledge is facilitated both by actively engaging them in analysis, reasoning, problem solving and debates with other students, and by two-way dialogs with the instructor during the class-wide discussions. This instructional approach also facilitates continuous formative assessment during instruction, especially in large classes, and makes students' thinking visible both to themselves and to their instructors, thereby allowing students to become defensive learners and permitting the instructor to address misconceptions as they arise. By presenting and evaluating qualitative arguments, students witness how principles and concepts are applied to solve problems, thereby helping students perceive and build an organized structure for their knowledge. Finally, our implementation of CCTs accommodates different learning styles and motivates even those students who are having difficulties, by showing that they are not the only ones having difficulties and that they can improve their understanding through perseverance.

The adoption of this approach is limited only by three factors: 1) Resources to buy and install the CCTs (CCTs are relatively inexpensive), 2) technical support and 3) faculty development to ensure that the appropriate pedagogy is used in teaching with this technology. This third point is crucial, since we have found that the occasional instructor who has attempted to use traditional instructional approaches with CCTs is less effective than if s/he had simply lectured. It is only when CCTs are used with an instructional approach that is more student-centered that it enhances learning. But this approach requires learning a new way to teach by faculty who often did well under the lecture method when they were students and who now have been lecturing for years.

Honors One-Credit Course on Teaching and Learning Science

Another course that has proved very popular among the students taking it has been a one-credit honors add-on course that is open only to those students taking either a physics or a biology CCT-taught course. The course, titled "Current Theories of Teaching and Learning and their Application to Science Instruction," enrolls about 20 students (ten from biology and ten from physics) and meets once a week for one hour. Each week a major topic in teaching and learning is the focus of discussion, with each class having a similar structure consisting generally of mini-lecture (ten minutes), group activity (20 minutes) and class-wide discussion (30 minutes). The topics covered and accompanying readings included: Reflections on Teaching and Learning[43]; constructivist epistemology[9,26,27]; preconceptions/misconceptions[9]; teaching for conceptual change[44-47]; knowledge organization in experts and novices[4,9]; learning styles[48,49]; National Science Education Standards[50]; Massachusetts State Frameworks for Science Education[51]; Cooperative Learning & Active Learning[52,53]; and innovations in physics instruction.[34,54-57]

There are no tests in the course. Weekly assignments consist of readings and keeping a journal. Journal writing assignments ask students to address specific issues discussed in the previous week's class (e.g., discuss how errors are different from misconceptions, or identify a misconception in biology or physics and discuss possible instructional strategies that would help students who held that misconception to overcome it). One major assignment is due at the end of the course that counts 20% of their grade and consists of critiquing either a high school or a college science class that they observed. The critiques, which are supposed to be based on all of what they had learned in the course, display a lot of learning and insights about the teaching and learning of science on the part of students. Overall grades in the course are a combination of attendance and participation in the weekly classes, the journals and the critique of the class they visited.

Students who enroll in this class are very interested in learning about teaching and learning, and could see a direct link between the research they reviewed and how their biology or physics class reflected the application of research findings to instruction. At the end of the course it is typical to find one or two students who decide to make teaching a career; others state that this course would prove very useful if they ever decide to enter the teaching profession.

A Course for Prospective High School Physics Teachers

The third course, titled "Motion, Interactions and Conservation Laws: An Active-Learning Approach to Physics" is specifically designed for undergraduates, graduates and inservice teachers interested in secondary physical science. Participants work with the NSF-funded Minds•On Physics (MOP) high school curriculum materials that were developed by our Physics Education Research group[58,59] in an activity-based mode to examine the following content areas: Motion, Forces, Dynamics, Newton's Laws, Momentum Conservation and Work-Energy. The course focuses on constructivist and active learning pedagogy useful in the teaching of physical science. In particular it examines ways to integrate hands-on activities with activities that promote concept development and problem-solving. The course makes extensive use of alternate representations (e.g., graphs and motion diagrams) and qualitative reasoning, and reviews the Massachusetts Science Frameworks as well as relevant science education research. In addition, participants develop activities and assessment techniques for use in teaching secondary physics.

Class time is spent in a combination of activities, including class-wide discussions, collaborative group work and modeling the type of coaching and support that should be provided to high school students. Discussion topics include physics content related to the MOP curriculum, pedagogy, assessment strategies and techniques, and educational research findings. The weekly homework assignments consist of reading and completing MOP activities, commenting upon physics education research articles,[60–62] and keeping a journal that was collected and graded periodically based on students' insights, views and experiences in the course.

Are Physics Departments Ready to Play a Major Role in the Preparation of K–12 Science Teachers?

The short answer to this question is "no," although there are several physics departments across the nation that are very actively involved in preparing K–12 science teachers. Several circumstances contribute to this situation. First, the training of PhD physicists at nearly all universities does not include courses on cognition, teaching and learning. In addition, teaching assistants are usually relegated to teaching traditional laboratory sections attached to traditionally taught large lecture courses. This situation is not conducive for developing pedagogical content knowledge, and so it is not surprising that young professors in physics departments do as was done to them and teach as they were taught. It is not easy to break out of this vicious cycle. The research on expertise clearly demonstrates that it takes time to become an expert at something, and becoming a competent "learning coach" is no exception. Thus, simply giving physics faculty "tips" in crash workshops on teaching and learning may serve to pique their interest, but does little to promote effective, or lasting, instructional innovations.

One solution to this problem that is promising for its versatility is for physics departments to invest in a physics education research (PER) group who would do scholarly work on teaching and learning physics, and who could educate graduate students and interested faculty on the subject of teaching and learning. PER faculty could design and implement an effective teaching and learning training component for graduate students so that upon graduation, young faculty are poised to break the "teach-as-you-were-taught" cycle. The PER group could also spearhead instructional innovations in service courses (including courses for preservice K–12 teachers) and in courses for physics majors. PER faculty could also play a major role in helping non-PER faculty in the department develop pedagogical content knowledge so that the latter could implement their own instructional innovations; this could be accomplished by working closely with interested faculty through the apprenticeship model, for example. A caveat here is not to try to force faculty to teach in a style with which they disagree or are not comfortable—if attempted, this will backfire, and it will be the students who suffer. It is best to enlist a core of dedicated faculty, preferably active researchers who command respect in the department, to apprentice with the PER group and with each other in making instruction a more meaningful experience for students.

There is also not a tradition in departments of physics to collaborate closely with faculty from schools of education. Such collaborations can prove fruitful in many regards. For example, a physics or a science methods course that is team-taught by a physics and an education faculty member would not only provide a rich educational experience for both faculty members but would also help students integrate physics and pedagogy, thereby helping students develop pedagogical content knowledge. Collaborations between physicists and educators can also help in crafting courses to meet the specific needs of preservice teachers and in devising ways of attracting more qualified students to the teaching profession.

From observing the dynamics at my own institution, it also helps tremendously to have an administration (department head, dean and provost) who actively support and reward teaching innovations. This serves to send a message throughout the institution that teaching is as important a mission as is research. It has also been my experience that there are numerous public relations benefits that accrue from a visible instructional reform movement at a university; it gives the administration ammunition to seek support from legislators, parents and trustees, many of whom nowadays feel that teaching is short-changed in research universities.

Another problem facing many research/flagship universities is that the number of students wishing to become physics or science teachers is small. Research/flagship universities attract some of the best prepared students in mathematics and science, and increasing the number of those students entering the teaching profession would decidedly benefit the state of science education in this country. Typical research universities have small numbers of physics majors, and it is often rare to find courses specifically designed for the needs of preservice teachers at all grade levels. By working closely with the school of education, physics departments could not only devise ways to increase the number of students wishing to become science teachers, but also design courses in collaboration with education faculty to address the specific needs of prospective teachers.

Discussion of Obstacles and Opportunities for Reform

In this article, I began by listing attributes that an "ideal" physics course for preservice science teachers should contain and went on to justify those attributes based on findings from learning research. I then described three courses in our Department that contain some of the attributes listed earlier. Finally, I described why the current infrastructure in physics departments is not optimal for playing a major role in the preparation of K–12 teachers. What should be evident from this article is that there are both obstacles and opportunities for changing the landscape in physics departments. I will conclude this article with a brief discussion of some of the obstacles and opportunities.

A major point I have tried to make in this article is that instructors at all levels need pedagogical content knowledge. An obstacle is that physics departments are not poised to help prospective teachers, graduate students, or their own faculty, acquire the necessary pedagogical content knowledge. On the other hand, opportunities abound. We now have a wealth of knowledge about learning and instruction that we can draw on in shaping reform. Two suggestions made in this article that could help overcome this "catch 22" obstacle is for physics departments to invest in a PER group and to nurture more collaborations with colleagues in schools of education.

Another obstacle is that the reward structure in research/flagship universities favors research over teaching. Although many of us can cite several cases where a faculty member earned tenure with strong research credentials and a mediocre teaching performance, finding cases of faculty who are tenured with a strong teaching case and a weak research case are extremely rare. Even tenured faculty who devote time and energy to scholarly work in educational reform are often viewed askance by colleagues and are not rewarded as well in the peer-review process that is common in academia. To foster opportunities for reform, we need to change the reward structure so that scholarship in teaching and learning is valued in research physics departments. I emphasize "scholarship" in the previous sentence since I do not simply mean that making cosmetic changes to a course (e.g., incorporating Web-based homework in a large course) should necessarily be considered scholarship in teaching and learning unless some important learning benefit is articulated and demonstrated (e.g., demonstrating that incorporating Web-based assignments that highlight the role of conceptual knowledge in problem solving results in increased "expert-like" behavior in students).

Another stubborn obstacle standing in the way of instructional reform is that faculty tend to teach the way they were taught, and the way they were taught is usually not optimal for the typical student and certainly not the way we would want preservice teachers to teach once they graduate. The opportunity here is that numerous physics curricula and instructional approaches now exist that have demonstrated their pedagogical value, so we do not need to continue doing what we have been doing since the invention of chalk. The catch is how to get faculty to use the curricula, or implement the instructional approaches the way they were intended to be used and implemented, since trying to retro-fit an innovation into a traditional setting often fails. The apprenticeship model, in which faculty understudy other faculty more experienced in implementing an innovation, has proved effective both at my institution and at the University of Illinois at Urbana/Champaign for creating a cadre of faculty who can rotate through important courses that are feeders for prospective teachers.

A number of obstacles fall somewhat outside the influence of academia. One is attracting more and better qualified prospective science teachers. Others falling in this category include teacher salaries, which are not all that competitive with the private sector; the lack of portability of retirement programs for teachers, which precludes their ability to change jobs after a few years at a school system; pay raises for teachers, unlike any other profession, are solely based on seniority and not merit, thereby stifling advancement and innovation; the lack of pedagogical content knowledge of the many teachers across the country who teach subjects outside their field of expertise; the heavy teaching loads of K–12 teachers precludes even teachers in the same school from discussing instructional innovations.

Despite these obstacles, we need to find ways at the national level of increasing the prestige, and elevating the importance, of the teaching profession. We are at an opportune time to make headway into both improving the number of preservice teachers and quality of science instruction they receive. To use an industrial analogy, to achieve these goals will require some "retooling" on the part of physics departments. Physics faculty need to invest some time in learning about research on learning and ways of applying findings from this research to teaching; the university administration, from the department level to the highest level, needs to support and reward scholarship into teaching and learning in the sciences; and the university's infrastructure needs to facilitate the blurring of the boundaries between education and the sciences.

*This paper is based on an invited talk presented at The Role of Physics Departments in Preparing K–12 Teachers conference held at the University of Nebraska on June 8–9, 2000. I would like to thank the University of Nebraska, the American Association of Physics Teachers, The American Physical Society and the American Institute of Physics for sponsoring the conference and commissioning this article. I would also like to thank Bob Dufresne, Bill Gerace,…, for providing valuable feedback on the manuscript.

References

1. Here "generic" does not imply a one-size-fits-all course for different levels (e.g., grades K–3 vs. 11–12) will likely look different to suit students' needs.
2. Redish, E.F., "Implications of cognitive studies for teaching physics," *American Journal of Physics*, **62**, 796–803, (1994).
3. Redish, E.F., "Discipline-based education and education research: The case of physics," *Journal of Applied Developmental Psychology*, **21**, 85–96 (2000).
4. Van Heuvelen, A., "Learning to think like a physicist: A review of research-based instructional strategies," *American Journal of Physics*, **59**, 891–897 (1991a).
5. McDermott, L.C., "Research on conceptual understanding in mechanics," *Physics Today*, **37** (7), 24–32 (1984).
6. McDermott, L.C., "Millikan Lecture 1990: What we teach and what is learned—closing the gap," *American Journal of Physics*, **59**, 301–315 (1991).
7. McDermott, L.C., "How we teach and how students learn—A mismatch?," *American Journal of Physics*, **61**, 295–298 (1993).
8. Mestre, J.P., "Learning and instruction in pre-college physical science," *Physics Today*, **44** (9), 56–62 (1991).
9. Mestre, J.P., "Cognitive aspects of learning and teaching science," in S.J. Fitzsimmons and L.C. Kerpelman (Eds.), *Teacher Enhancement for Elementary and Secondary Science and Mathematics: Status, Issues and Problems* (pp. 3-1–3-53). Washington, DC: National Science Foundation (NSF 94-80), February 1994.
10. Mestre, J., and J. Touger, "Cognitive research: What's in it for physics teachers," *The Physics Teacher*, **27** (September), 447–456 (1989).
11. National Research Council, *How People Learn: Brain, Mind, Experience, and School*, Washington, DC: National Academy Press, 1999.
12. Chi, M.T.H., and R. Glaser, "The measurement of expertise: Analysis of the development of knowledge and skills as a basis for assessing achievement," in E.L. Baker and E.S. Quellmalz (Eds.), *Design, Analysis and Policy in Testing* (pp. 37–47), Beverly Hills, CA: Sage Publications, 1981.

13. Glaser, R., "Expert knowledge and processes of thinking," in D. Halpern (Ed), *Enhancing Thinking Skills in the Sciences and Mathematics* (pp. 63–75), Hillsdale, NJ: Lawrence Erlbaum Associates, 1992.
14. Larkin, J.H., "Information processing models in science instruction," in J. Lochhead and J. Clement (Eds.), *Cognitive Process Instruction* (pp. 109–118), Hillsdale, NJ: Lawrence Erlbaum Assoc., 1979.
15. Chi, M.T.H., P.J. Feltovich and R. Glaser, "Categorization and representation of physics problems by experts and novices," *Cognitive Science*, **5**, 121–152 (1981).
16. Dufresne, R., W.J. Gerace, P.T. Hardiman, and J.P. Mestre, "Constraining novices to perform expert-like problem analyses: Effects on schema acquisition," *Journal of the Learning Sciences*, **2**, 307–331 (1992).
17. Eylon, B.S., and F. Reif, "Effects of knowledge organization on task performance," *Cognition & Instruction*, **1**, 5–44 (1984).
18. Heller, J.I., and F. Reif, "Prescribing effective human problem solving processes: Problem description in physics," *Cognition and Instruction*, **1**, 177–216 (1984).
19. Leonard, W.J., R.J. Dufresne, and J.P. Mestre, "Using qualitative problem-solving strategies to highlight the role of conceptual knowledge in solving problems," *American Journal of Physics*, **64**, 1495–1503 (1996).
20. Mestre, J.P., R. Dufresne, W.J. Gerace, P.T. Hardiman and J.S. Touger, "Promoting skilled problem solving behavior among beginning physics students," *Journal of Research in Science Teaching*, **30**, 303–317 (1993).
21. Mestre, J., R. Dufresne, W. Gerace, P. Hardiman and J. Touger, "Enhancing higher-order thinking skills in physics," in D. Halpern (Ed.), *Enhancing Thinking Skills in the Sciences and Mathematics* (pp. 77–94), Hillsdale, NJ: Lawrence Erlbaum Associates, 1992.
22. Chapman, O.L., "Learning science involves language, experience, and modeling," *Journal of Applied Developmental Psychology*, **21**, 97–108 (2000).
23. Anderson, C.W., "Strategic teaching in science," in B.F. Jones, A.S. Palincsar, D.S. Ogle and E.G. Carr (Eds.), *Strategic Teaching and Learning: Cognitive Instruction in the Content Areas* (pp. 73–91), Alexandria, VA: Association for Supervision and Curriculum Development, 1987.
24. Schauble, L., "Belief revision in children: The role of prior knowledge and strategies for generating evidence," *Journal of Experimental Child Psychology*, **49**, 31–57 (1990).
25. Resnick, L.B., "Mathematics and science learning: A new conception," *Science*, **220**, 477–478 (1983).
26. Glasersfeld, E., "Cognition, construction of knowledge, and teaching," *Synthese*, **80**, 121–140 (1989).
27. Glasersfeld, E., "A constructivist's view of learning and teaching," in R. Duit, F. Goldberg and H. Niedderer (Eds.), *The Proceedings of the International Workshop on Research in Physics Education: Theoretical Issues and Empirical Studies* (Bremen, Germany, March 5–8, 1991), Kiel, Germany: IPN, 1992.
28. Vosniadou, S., and W.F. Brewer, "Mental models of the earth: A study of conceptual change in childhood," *Cognitive Psychology*, **24**, 535–585 (1992).
29. Shulman, L., "Those who understand: Knowledge growth in teaching," *Educational Researcher*, **15** (2), 4–14 (1986).
30. Shulman, L., "Knowledge and teaching: Foundations of the new reform," *Harvard Educational Review*, **57**, 1–22 (1987).
31. Minstrell, J.A., "Teaching science for understanding," in L.B. Resnick and L.E. Klopfer (Eds), *Toward the Thinking Curriculum: Current Cognitive Research* (pp.129–149), Alexandria, VA: Association for Supervision and Curriculum Development, 1989.
32. Dufresne, R.J., W.J. Gerace, R.J. Leonard, J.P. Mestre, J.P. and L. Wenk, "Classtalk: A classroom communication system for active learning," *Journal of Computing in Higher Education*, **7**, 3–47 (1996).

33. Mestre, J.P., W.J. Gerace, R.J. Dufresne and R.J. Leonard, "Promoting active learning in large classes using a classroom communication system," in E.F. Redish and J.S. Rigden (Eds.), *The Changing Role of Physics Departments in Modern Universities: Proceedings of the International Conference on Undergraduate Physics Education/Part Two: Sample Classes* (pp. 1019–1036), Woodbury, NY: American Institute of Physics, 1997.

34. Wenk, L., R. Dufresne, W. Gerace, W. Leonard and J. Mestre, "Technology-assisted active learning in large lectures," in A.P. McNeal and C. D'Avanzo (Eds.), *Student-active science: Models of innovation in college science teaching* (pp. 431–451), Orlando, FL: Saunders College Publishing, 1997.

35. Laws. P., "Calculus-based physics without lectures," *Physics Today*, **44** (12), 24–31 (1991).

36. McDermott, L.C., & the Physics Education Group at the University of Washington, *Physics by Inquiry*, New York: Wiley, 1996.

37. Mestre, J.P., "Progress in research: The interplay among theory, research questions, and measurement techniques," in A.E. Kelly and R.A. Lesh (Eds.), *Handbook of Research Design in Mathematics and Science Education* (pp. 151–168), Mahwah, NJ: Lawrence Erlbaum Associates, 2000.

38. A concept scenario is a sequence of concepts that apply to a problem in the order in which they apply. For example: Mechanical energy is conserved, followed by conservation of momentum, followed by conservation of mechanical energy, with potential energy increasing and kinetic energy decreasing.

39. Brown, A.L., "The development of memory: Knowing, knowing about knowing, and knowing how to know," in H.W. Reese (Ed.), *Advances in Child Development and Behavior, Vol. 10*, New York: Academic Press, 1975.

40. Brown, A.L., "Metacognitive development and reading," in R.J. Spiro, B.C. Bruce and W.F. Brewer (Eds.), *Perspectives from Cognitive Psychology, Linguistics, Artificial Intelligence, and Education*, Hillsdale, NJ: Lawrence Erlbaum Associates, 1980.

41. Flavell, J.H., "Metacognitive aspects of problem solving," in L.B. Resnick (Ed.), *The Nature of Intelligence*, Hillsdale, NJ: Lawrence Erlbaum Associates, 1973.

42. Bransford, J., S. Brophy and S. Williams, "When computer technologies meet the learning sciences: Issues and Opportunities," *Journal of Applied Developmental Psychology*, **21**, 59–84 (2000).

43. Duckworth, E., "The having of wonderful ideas" and other essays on teaching and learning. New York: Teachers College Press, 1987.

44. Brown, D., "Using examples to remediate misconceptions in physics: Factors influencing conceptual change," *Journal of Research in Science Teaching*, **29**, 17–34 (1992).

45. Brown, D., and J. Clement, "Overcoming misconceptions via analogical reasoning: Factors influencing understanding in a teaching experiment," *Instructional Science*, **18**, 237–261 (1989).

46. Clement, J., "Using bridging analogies and anchoring intuitions to deal with students' preconceptions in physics," *Journal of Research in Science Teaching*, **30** (10), 1241–1257 (1993).

47. Posner, G., K. Strike, P. Hewson and W. Gerzog, "Accommodation of a scientific conception: Toward a theory of conceptual change," *Science Education*, **66**, 211–227 (1982).

48. Svinicki, M.D., and N.M. Dixon, "The Kolb Model modified for classroom activities," *College Teaching*, **35**, 141–146 (1987).

49. Felder, R.M., and L.K. Silverman, "Learning and teaching styles in engineering education," *Engineering Education*, **78** (April), 674–681 (1988).

50. National Research Council, *National Science Education Standards*, Washington, DC: National Academy Press, 1996.

??51. Massachusetts State Frameworks for Science Education.

52. Johnson, D.W., R.T. Johnson and K.A. Smith, *Active Learning: Cooperation in the College Classroom*, Edina, MN: Interaction Book Co., 1991.

53. Sutherland, T.E., and C.C. Bonwell, "Using active learning in college classes: A range of options for faculty," *New Directions for Teaching and Learning*, San Francisco, CA: Jossey-Bass, 1996.

54. Heller, P., R. Keith and S. Anderson, "Teaching problem solving through cooperative grouping. Part 1: Group versus individual problem solving," *American Journal of Physics*, **60**, 627–636 (1992).

55. Heller, P., and M. Hollabaugh, "Teaching problem solving through cooperative grouping. Part 2: Designing problems and structuring groups," *American Journal of Physics*, **60**, 637–644 (1992).

??56. Mazur, 1997.

57. Van Heuvelen, A. "Overview, case study physics," *American Journal of Physics,* **59**, 898–907 (1991b).

58. Leonard, W., R. Dufresne, W. Gerace and J. Mestre, J. *Minds•On Physics* (Vol. 1: Motion; Vol. 2: Interactions; Vol. 3: Conservation Laws & Concept-Based Problem Solving, Vol. 4·FF: Fundamental Forces & Fields, Vol. 4·AT: Advanced Topics in Mechanics), Dubuque, IA: Kendall/Hunt Publishing, 1018 pp. (1998, 1999).

59. Leonard, W., R. Dufresne, W. Gerace, and J. Mestre, Teacher's Guides to accompany *Minds•On Physics* (4 volumes), Dubuque, IA: Kendall/Hunt Publishing, 1564 pp. (1998, 1999).

60. Hestenes, D., M. Wells and G. Swackhamer, "Force concept inventory," *The Physics Teacher*, **30** (March), 159–166 (1992).

61. Hestenes, D., and I. Halloun, "Interpreting the force concept inventory: A response," *The Physics Teacher*, **33** (November), 502, 504–506 (1995).

62. Huffman, D., and P. Heller, "What does the force concept inventory actually measure?" *The Physics Teacher*, **33**, (March), 138–143 (1995).

63. Gick, M.L., and K.J. Holyoak, "Analogical problem solving," *Cognitive Psychology*, **12**, 306–355 (1980).

??. Gerace, W.J., W.J. Leonard and R.J. Dufresne, ""Using technology to implement active learning in large classes," in G. Sastre (Ed.), *Constructivism and New Paradigms in Science and Education*, Buenos Aires: Gedisa, in press.

Figure 1.
The difficulty of transferring knowledge flexibly across contexts.

A study conducted by Gick and Holyoak[63] asked college students to memorize the following passage about a general trying to take over a fortress:

A general wishes to capture a fortress located in the center of a country. There are many roads radiating outward from the fortress. All have been mined so that while small groups of men can pass over the roads safely, a large force will detonate the mines. A full-scale attack is therefore impossible. The general's solution is to divide his army into small groups, and send each group to the head of a different road, and have the groups converge simultaneously on the fortress.

After students showed that they had assimilated this passage, they were given the following problem to solve:

You are a doctor faced with a patient who has a malignant tumor in the stomach. It is impossible to operate on the patient, but unless the tumor is destroyed the patient will die. There is a kind of ray that may be used to destroy the tumor. If the rays reach the tumor all at once and with sufficient high intensity, the tumor will be destroyed, but surrounding tissue may be damaged as well. At lower intensities, the rays are harmless to healthy tissue, but they will not affect the tumor either. What type of procedure might be used to destroy the tumor with the rays, and at the same time avoid destroying the healthy tissue?

Although few students could solve this problem on their own, when they were asked to use the information from the previous passage, over 90% of them were able to solve the problem. Despite the parallelisms between the two stories, students did not spontaneously use the information from the passage of the general and the fortress to solve the tumor problem.

Re-Preparing the Secondary Physics Teacher

Fred Stein
American Physical Society

Introduction

This case study describes a successful and on-going program begun at Colorado State University (CSU) in 1994. It dramatically improved the preparation of secondary science and mathematics teachers through the revision of targeted science, mathematics and education courses, and the expansion and enrichment of field experiences.

From 1994–1999, CSU was the lead institution in a project called the Rocky Mountain Teacher Education Collaborative, one of the sites of the NSF-supported Collaboratives for Excellence in Teacher Preparation. CSU worked in concert with two other universities and three community colleges (University of Northern Colorado, Metropolitan State College of Denver, Community College of Denver, Aims Community College and Front Range Community College). The program continues to impact all prospective science and mathematics teachers at CSU (including chemistry, geology, biology and mathematics teachers) and many faculty members. Key to the success of the programs was the institutionalization of the following program elements. (1) A long-term, active collaboration between the science and mathematics departments and the CSU school of education; (2) A productive Teacher-in-Residence program that is used as a "reality check" for reform; (3) The redesign of content and pedagogy for targeted science and mathematics courses based on results from science and mathematics education research; (4) The integration of learning theory, teaching methods, and science and mathematics content through a revised and team-taught science methods course; and (5) The participation of science and mathematics faculty in the assignment and supervision of practicum and student-teaching experiences.

Background and Goal

From 1984, beginning with *A Nation at Risk*, through 1996, with *Shaping the Future*[1] and recently with *To Touch the Future: Transforming the Way Teachers are Taught*,[2] national reports have decried the inadequate preparation and lack of competency of new science teachers at all levels K–12. If it is true that "teachers teach as they were taught," then the vision for improving physical science and physics teaching and learning in K–12 should be that universities model effective teaching/learning approaches in their science courses for both majors and non-majors. This was the vision that drove the program discussed in this case study. The four-year institutions in this project encouraged science and mathematics departments to collaborate with their departments of education to plan and improve the science preparation of future secondary science teachers through the revision of targeted courses and enriched field experiences. Ultimately, the goal was to produce better-prepared science and mathematics teachers who were committed to the objectives of the national reform movements such as the National Science Education Standards (NRC), the Benchmarks from Project 2061 (AAAS) and the three volumes of Standards from the National Council of Teachers of Mathematics (NCTM).

Project Activities
Over five years, the following components were put in place:
- A long-term, active collaboration between the physics, chemistry, biology, geology and mathematics departments, the school of education, the local two-year college and the local school community
- A productive Teacher-in-Residence (TIR) program that was used as a "reality check" for reform
- The redesign of content and pedagogy for targeted science and mathematics courses by faculty based on results from science and mathematics education research and utilizing appropriate technologies
- The integration of learning theory, teaching methods and science content through the revised and team-taught science and mathematics methods courses
- The participation of science and mathematics faculty in the assignment and supervision of practicum and student teaching experiences.

Project Components
Teachers in Residence
The TIR brought the knowledge and experience of managing a student-centered science class, assisted faculty in revising targeted science and mathematics courses and helped team-teach the science and mathematics methods courses. The TIR at CSU was housed in the Center for Science, Mathematics & Technology Education (CSMATE) facility but also worked with the school of education and the local schools. The TIR provided continuity between the science and mathematics methods courses, the science courses and the activities in the local schools. The university reimbursed the school district for the cost of the TIR's replacement. Later, CSMATE provided the TIR with a mini-grant for classroom supplies when

they returned to the classroom. The TIR consulted with preservice teachers and provided a realistic understanding of what is being taught and how it is being taught in the schools. The TIR offered valuable contacts with local teachers and school districts that significantly improved practicum activities and the placement of student teachers.

Professor Sanford Kern reflects on his experience working with TIR:

"Two Teachers-in-Residence were involved in the physics classes, one each during the 1996/97 and 1997/98 academic years. Their presence was invaluable. They were able to communicate directly to me about the backgrounds, knowledge base, and likely responses of students coming from high school. For their part, they gained insight into what we as a university community expect from students in our classes. The two Teachers-in-Residence differed in their contributions to the class, and in what they took away. One played a very active and direct role in teaching the class and brought in many materials to use, such as videos, which were not standard fare at Colorado State University. The other was less active directly but devised 50 or more activities that could be used in both high school and university settings. Both appreciated the professional development work with physics faculty members while increasing their depth of understanding and updating their content knowledge. They both helped me understand how the K–12 State Standards are applicable to physics courses, and at the same time then, saw the latest content and areas of emphasis at the university level. They were glad to develop new relationships and alliances with university personnel. In summary the two Teachers-in-Residence who were involved with my physics classes returned to the public school system with a greater sense of satisfaction and dedication to the development of relationships between K–12 teachers and higher education faculty. Upon returning to the public school system, they provided professional development training to other teachers on new content updates based on the research they observed while working with higher education faculty, as well as ideas on how to better link K–12 courses to introductory physics courses at the university. They also served as cooperating teachers for our student teachers, thus promoting the ideals of the reformed classroom. They leave a legacy of improved instructional strategies within undergraduate classrooms, and continue to provide a network at the local level for K–12 reform."

Targeted Science and Mathematics Courses

Central to the project's activities was the restructuring of targeted courses and their instructional approaches. The appropriate sections of the first-year introductory science and mathematics courses were redeveloped to promote active learning, preferably in an integrated lecture and laboratory format. The redesigned courses encouraged less reliance on the authoritarian, teacher-dominated, transfer model of science instruction and require a more spontaneous interchange of ideas. Also, the ubiquitous availability of laboratory equipment enabled students to discover relationships, as well as confirm them. Other changes in instructional strategies included bringing more inquiry-based, student-centered experiences into lecture sections through cooperative learning and peer-coaching techniques, enhancing learning using technology and other successful delivery systems aimed at actively engaging students.

Again, Professor Kern reflects:

"*Introductory courses pH 121 and pH 122 are algebra-based with a laboratory component, and students who take them come from a wide variety of disciplines, with many from the biological sciences. The courses are recommended for preservice science teachers, although some prospective physics and mathematics teachers take a calculus-based course. Because students who enroll in the pH 121/122 series come from very diverse backgrounds, often with little prior exposure to physics, these courses require the greatest degree of in-class explanation, demonstration, and motivation. They also are appropriate for modeling behaviors we wish future teachers to adopt—namely, an integrated lecture-laboratory approach based on understanding the material and problem solving, rather than a more traditional approach emphasizing rote memorization and application. This paper will further describe some observations about teaching an integrated lecture, laboratory, and recitation class in an 'Experiential Learning Studio' environment.*"

Classes met for 2 1/2 hours, twice a week. The classroom was arranged with 24 desks clustered in groups of four, each forming an octagon with students seated on the inside. Students had their own space, yet there was an easy flow to interactions among members of each cluster and free flow for the instructor as well. Students took notes, solved in-class problems, performed laboratory experiments, and took exams at these desks, often acting cooperatively with their 'clustermates.' While the composition of clusters occasionally changed, in general they tended to be stable, with some groups continuing to operate more than two years after the class ended, and even maintaining contact after some members left campus. A good deal of camaraderie resulted from this use of cooperative learning, which promoted teamwork and 'information sharing.' Most class sessions consisted of a variety of activities, including a mixture of problem-solving demonstrations, and questions about homework from which further in-class questions developed—some emanating from students, many from the instructor. It was common to have a series of questions posed, escalating in difficulty and depth of understanding. As the instructor, I was not at all reluctant to make an occasional class quite intense. During these sessions, attention was highly focused and students gained a great deal from the concentrated emphasis on concepts and immediate application to practical problems they were asked to comment on and solve.

From student feedback, we can see the importance of addressing the kinesthetic component of learning for those students who have had less experience than others. The more closely students can tie experience to concepts or theory the greater their understanding of the subject matter. Importantly this models the in-class behavior we wish preservice science teachers to adopt for their core teaching. Facilitating this in a cooperative and participatory-environment brings us closer to approaching true facilitation of learning. Student response was overwhelmingly positive to the new initiatives. Over 80 percent of the class thought that having enough class time to answer questions was 'very-to-extremely' helpful, and 95 percent also agreed that the course was intellectually challenging."

Science-Methods Courses

General Methods of Teaching and Science Methods were combined into one two-semester course and was team-taught by science and mathematics faculty, education faculty and the TIR. The course utilized innovative (and tested) teaching methods, which are student-centered and inquiry-based. Including in-depth discussions of state and national content Standards, techniques for effective classroom management, skills in curriculum development and planning, and experience in inquiry-based and constructivist theory and practice in teaching and learning (constructivist methods emphasize the active engagement of the student in the learning process and recognizes the importance of prior knowledge for new learning). Hands-on experiences followed by structured reflection about what the student observed and what can be inferred and the infusion of science content was a significant part of the course. This course enabled preservice teachers to teach their future students to do science, which includes how to encourage scientific habits of mind. Integrated into this course was an on-going study of issues relating to equity and diversity. Students were expected to understand and practice research-based strategies for overcoming the educational barriers experienced by women and minority students in their study of science.

Field-Based Experiences

The quality of field experiences of future teachers was a central concern of the project. Contact between the campus program and teachers in the local schools were strengthened to establish a cadre of reform-minded cooperating teachers who would receive and help prepare the preservice teachers. Initially, outstanding teachers and later former TIRs served as cooperating teachers for new student teachers whenever possible. Ideal practices included following two years of supervised practicum activities (observations and classroom aides) in local K–12 schools with monthly seminars given by the supervising team (the education specialist and the science or mathematics faculty). The student had occasional school visits by faculty during the first year of employment (the induction year) for in-school coaching, evaluating, mentoring and other supportive activities.

Evaluation

An external evaluation team was hired to provide evaluation in the form of accountability (formative implementation evaluation), feedback (formative progress evaluation) and outcomes (summative evaluation). To provide evidence of how well the goals of the project have been realized, the external evaluation team determined the answers to certain effectiveness questions concerning each of the components:

- How effective and long-term was the collaboration between the science and mathematics departments and the schools of education at CSU and the other sites?
- How many courses were redesigned, revised and institutionalized at each site?
- How many faculty were involved during the project's funding and how many continued to be involved after the external support was completed?
- What significant unintended impacts did the project have on the university environment?
- How many teachers received their degree/ certification? Where did they go?
- Did the project make a significant difference in the teaching skills of the graduates?
- Was there a positive difference in the students' attitudes and understandings in science and mathematics because of the graduates' training?

Professor Kern attempted to quantify his evaluation:

"The value of utilizing new strategies and methods, however, is measured by how well students learn. The first exam given to the initial class of 23 students was identical to the one given to the large, traditional lecture section of 485 students. My section grades averaged 60 percent, compared with the large section's 55 percent. These results were gratifying, since they indicated the populations were close to equivalent. On the next test, we decided to go to a nontraditional method of assessment, rather than using the multiple-choice method that was used for the first examination. However, we did include two questions on the second examination for our class that was used by the traditional section. The traditional section averaged 82 percent and 62 percent, compared with our class section's average of 96 percent for the two questions. We concluded that the conceptual emphasis used in our class did not pose a barrier to solving 'normal' types of questions. During the course of the year we could see C- and even D+ students performing

at a C+ and B- level. Students attributed some of their increased understanding to doing 'hands-on' work as they were exploring and discussing concepts to actually seeing and knowing physics. We saw each cluster sharing and interchanging responsibilities. Gratifyingly, female students actively participated, and often assumed leadership roles. The structure also allowed team members to respond quickly to errors."

The New Project—PhysTEC

Recently, The American Physical Society (APS), in partnership with the American Association of Physics Teachers (AAPT) and the American Institute of Physics (AIP), submitted a proposal to the National Science Foundation (NSF) and to the Fund for the Improvement of Postsecondary Education (FIPSE) to dramatically improve the preparation of physics and physical science teachers nationwide. This project will increase the role of physics departments, in collaboration with education departments, to create more and better-prepared future teachers. Over the next five years, NSF/FIPSE will enable the Physics Teacher Education Coalition (PhysTEC) to be established with an initial membership of more than 25 universities and colleges that share an increasing interest in revising their program in teacher preparation.

The Coalition, through the leadership and assistance of the physics professional societies, will provide professional development for participating faculty and a vehicle for dissemination and research. PhysTEC will build on the experiences at CSU and other exemplary models that have documented their best practices. Although PhysTEC builds on components of the CSU experience, it transforms and inverts the strategy from a focus on several science and mathematics disciplines at a single geographical site to that of the nationwide reform of a single discipline (physics) aimed at a large number of college and university sites.

The Coalition will consist of three levels of institutional involvement. Six to eight of the institutions within the coalition will be selected as "Primary Program Institutions" (PPIs) and will be substantially supported by other external funders. They will engage in creating, describing and evaluating a set of model programs that are flexible in nature but will undertake some specific elements of programmatic change. Other schools with significant track records in teacher preparation will serve as "Resource Institutions" (RIs) that will be supported to consult and offer professional development advice and experience to the PPIs. The last group of institutions has demonstrated interest but is presently not involved in teacher-preparation programs. The involvement of the professional societies will make the coalition possible by providing access to the members of the broad physical-science community through the APS/AAPT leadership, committees, national meetings and conferences, workshops, and publications, and to those who can influence policy regarding teacher preparation.

APS/AAPT/AIP have identified preservice (prospective) teacher preparation as a key issue for the physics community. To this purpose the three organizations recently approved a joint statement in which they "urge the physics community, specifically physical science and engineering departments and their faculty members, to take an active role in improving the preservice training of physics/science teachers." These societies now recognize a responsibility to assist physics departments and their faculty in developing strategies and implementing changes that allow them to constructively respond to this challenge.

As the focusing effort of PhysTEC, each of the PPIs will develop and implement a model program of study uniquely suited to their institution. The models will be informed by successful elements from previous projects emphasizing science and mathematics preparation for teachers. The PPIs will, in collaboration with PhysTEC staff members and each other, establish and publish guidelines to facilitate other institutions' adapting, implementing and evaluating these models. The PPIs, as models, will be expected to implement each of the following program elements in a productive manner that best suits their institutional environment.

- A long-term, active collaboration between the physics department, the school or department of education, the local two-year colleges and the local school community
- A productive TIR program as a "reality check" for reform
- The redesign of content and pedagogy for targeted physics courses by physics faculty based on results from physics education research and utilizing appropriate interactive technologies
- The integration of learning theory, teaching methods and physics content through revised and team-taught education science-methods courses
- The participation of physics faculty in the assignment and supervision of practicum and student teaching experiences

The selection of PhysTEC PPIs will be based on the individual institution's
1. commitment to become actively involved with teacher preparation reform as demonstrated by their degree of effectiveness (success) with previous efforts,
2. readiness to work in collaboration with faculty from the school of education,
3. degree of enthusiasm to model good teaching practices and their capacity to document their work and serve as a model for others within the higher education community,
4. willingness to shift some of their own resources toward PhysTEC and
5. capability for program institutionalization.

Joining the efforts of APS/AAPT/AIP provides a unique opportunity for leadership, consultation and technical assistance through a single discipline that can encourage collaborations between departments of physics and education to implement and refine changes. There is preliminary evidence that an effort to improve the physics and physical-science education of elementary and secondary preservice teachers can contribute to the general revitalization of undergraduate science education. Physics departments that focus on providing good models in their introductory courses for preservice teachers will learn to adapt their efforts to improve the teaching and learning for all students.

During the academic year, teams of physics and education faculty (including department heads) from the PPIs will visit the campuses of the RIs. During the following summer, a conference and workshops will be held for all participants (plus newly interested institutions) led by the RI faculty and the PhysTEC staff to share, reinforce and develop strategic plans for the project implementation phase. Then PPIs will be expected to implement and institutionalize the program elements in a manner that best suits their institutional environment.

References

1. National Science Foundation, 1996.
2. ACE, 1999.

Keynote Address

THE ROLE OF UNIVERSITY PRESIDENTS IN PREPARING TEACHERS

I appreciate the opportunity to meet with you today and to address one of the issues that I consider critical for the future of our nation—how to improve education at the K–12 level, especially in the sciences.

As some of you know, I served on the American Council on Education's Presidents' Task Force on Teacher Education, which last year issued its report, "To Touch the Future: Transforming the Way Teachers are Taught" (American Council on Education, 1999). I have also been active, along with my colleague Dr. Doug Christensen, Nebraska's Commissioner of Education, and others, in what we've come to call the P–16 Initiative. The P–16 Initiative is an effort to establish a coordinated system of education from preschool through college—a system that demands high standards for students and for teachers at every level of education. Nebraska is not alone in such efforts. About 20 states are now engaged in some significant program to assure high quality in a seamless system of education from the primary level through the baccalaureate.

What I propose to do, in the next few minutes, is to look at some teacher education issues from the vantage point of the ACE Presidents' Task Force. Then, I would like to describe briefly some of the work we've accomplished here in Nebraska on the P–16 Initiative.

The ACE report was directed to presidents of America's colleges and universities, and it sets forth an action plan describing steps presidents must take to improve teacher education in the nation. The report emphasizes two key points: 1) access to high-quality education ranks at the top of concerns among the American people; and 2) teachers exert a singularly powerful influence on the academic performance of students.

The public expects colleges and universities to produce teachers who are knowledgeable about what they teach and effective in how they teach. This means teachers must have a thorough grounding in the subject matter they are teaching and they must be competent in pedagogy as well. Recent studies using refined methods to measure teacher qualifications have established that the number and kind of courses taken by math and science teachers do, in fact, influence student performance. Students learn more math and science when their teachers have taken college-level math and science courses. The more courses—and more advanced the courses—the teacher has taken, the better the students perform. Evidence of such clear linkages between teacher preparation and student performance has not been established in all subject matter areas. But it seems reasonable to assume the results of other such studies would parallel those in math and science.

There is a serious public policy issue at stake here: Many school systems employ teachers who are under-prepared. According to Kati Haycock, director of the Education Trust, nearly 25 percent of the nation's secondary school courses—and more than 33 percent of the courses in predominantly minority schools—are taught by teachers who do not have even a college minor in the subjects they are teaching. But Haycock adds that even teachers with college minors or majors in the fields they are teaching often struggle with content when they are trying to move their students to deeper levels of understanding.

The Education Trust points out that seven states have no licensing exams at all for new teachers. Forty-four states require that prospective teachers pass exams, but only 29 states require that secondary school teachers pass an exam in the subject area in which they plan to teach. In those states where subject-matter testing is required, the knowledge required to pass is high-school level. None of the tests examined by the Education Trust required knowledge at the baccalaureate level or above. Obviously, this is not good enough. No one would accept treatment from a health-care professional with such poor evidence of qualification. We cannot have our children and grandchildren in classrooms with unqualified teachers. The ACE report calls on presidents to become leaders in efforts to eliminate this situation. The fact that many students in America—often those most in need of good teachers—are taught by unqualified teachers is reprehensible.

The good news is that states are moving to help change the situation. A number of performance-based licensure requirements are emerging nationwide, using such methods as examination portfolios and videotaped teaching sessions. Internships are being instituted that parallel clinical licensure procedures in the health-care field. And some states are requiring standardized assessments of classroom practice and student work samples. Obviously, universities and colleges cannot solve these problems unilaterally, but must work closely with state education departments with school systems.

We can—and must—however, move the education of teachers to the top of our agenda, making it clear that teacher education is strategically important to the mission of our institutions. We can make standards higher; both for entrance into education colleges and graduation from them. We can conduct serious reviews of the quality of education we provide prospective teachers across all our colleges.

The ACE Task Force commissioned a study to identify characteristics of a successful program for improvement of teacher education. They include:

- Development of an effective way to combine the contributions of the arts and science faculty and the education faculty
- Support from the president's office and from state and community school leaders

A thoughtfully designed process for admitting students to teacher education programs

Subject matter and clinical training that are closely articulated, coupling theory and practice

Program outcomes that are carefully, independently, and continuously assessed

Attentive guidance by university faculty and school leaders for new teachers following graduation so that they are not just set adrift

Related to this last point, "clinical partnerships" between departments and the schools could provide new teachers with continuing assistance and mentoring following graduation. Such partnerships could enhance significantly the success rate of new teachers. Just as important, departments could work in partnership with the schools to assist both new and experienced teachers with professional development opportunities that use both the faculty and the research resources of the university. These partnerships would help teachers stay current in subject matter knowledge and the use of new technologies. They would also allow teachers to maintain an essential collegial link with college and departmental faculty at the university.

To accomplish such partnerships, we will have to overcome a common problem in colleges and universities—the tendency for faculty and programs to become isolated within departmental boundaries. We must keep in mind that, at both the secondary and elementary levels, science teachers may be teaching more than one of the sciences. It is important that we develop cooperative interdepartmental courses and professional development training for these teachers. Responsibility for preparing prospective teachers in subject matter areas rests with both the college of education faculty and the faculty within the departments of the colleges that instruct prospective teachers—particularly faculty in the arts and sciences.

Let me turn now, for a few minutes, to the work we've done here in Nebraska on the so-called P–16 Initiative. Commissioner Doug Christensen of the Nebraska Department of Education and I began working on this program in 1997. This followed some national meetings at which Nebraska educators heard reports on efforts in other states, such as Maryland, to coordinate education in a continuum from kindergarten through college.

One result was our agreement on principles developed in these national meetings, specifically that we will ensure that all high-school graduates meet high standards, that we will accept only teachers who can bring student performance to high standards, that we will accept into college only students who meet high standards, and that we will ensure that all teachers candidates we produce are prepared to bring student performance to high standards.

Another outcome of these meetings was the development of a compact signed by a number of colleges of education in the state—including the three at the University of Nebraska—with the following provisions:

> We will assess the subject-area knowledge of our graduates to demonstrate that they have sufficient expertise to teach students to meet national and state standards using the best available test.
>
> We will retain experienced P–12 teachers to serve as independent evaluators of the teaching skills of our students at the time of graduation.
>
> We will ask schools to report on the academic progress of the students taught by our graduates. This will include the test scores currently used by schools to assess academic achievement.
>
> We will ask the administrative supervisors of our graduates to evaluate their effectiveness as teachers.
>
> We will ask our graduates to evaluate how well the college prepared them for the realities of the classrooms in which they teach.

All of the information resulting from the provisions of this compact will be made public so that people can make informed judgment about teacher education programs. And, of course, the information will be used to make continuous improvements to teacher education programs.

Several meetings of the State Board of Education and the University's Board of Regents were held subsequently, culminating in resolutions adopted in the fall of last year by each body. These resolutions supported the idea of a partnership to work toward alignment of K–12 academic standards with admissions standards for higher education institutions in the state. Based on those resolutions, a steering committee has been formed, and a P–16 Council made up of representatives of both private and public education at all levels has begun a series of meetings to determine specific action steps that will be required.

While the coordination and alignment of the various educational systems involved—together with the demand for high standards—is complex, we are confident that we will be successful in significantly upgrading education in this state at all levels.

L. Dennis Smith, President, University of Nebraska

Acknowledgements

Many individuals have contributed to this initiative through their vision, implementation efforts, and/or funding support. They are (alphabetically):

University of Nebraska
Royce Ballinger, Director, Nebraska EPSCoR
Gayle Buck, Assistant Professor, Department of Curriculum & Instruction and Conference Co-Chair
Patricia Christen, Business Manager, Department of Physics & Astronomy
Vicki Plano Clark, Laboratory Manager, Department of Physics & Astronomy
Brian Foster, Former Dean of the College of Arts & Sciences
Elizabeth Franklin, Chair, Department of Curriculum & Development
Roger Kirby, Chair, Department of Physics & Astronomy
Diandra Leslie-Pelecky, Assistant Professor, Department of Physics & Astronomy and Conference Co-Chair
Jim Lewis, Professor, Department of Math & Statistics and Chair, Math/Science Education Initiative
Marilyn McDowell, Conference Secretary
Linda Pratt, Interim Dean of the College of Arts & Sciences
Sandra Scofield, Director, Center for Math, Science & Computer Education
L. Dennis Smith, President, University of Nebraska System
James O'Hanlon, Dean of Teachers College
Laura White, Assistant Dean of the College of Arts & Sciences

Associates
Robert C. Hilborn, Professor of Physics, Amherst College and Chair, National Task Force on Undergraduate Physics
Suzanne Kirby, Teacher, Clinton Elementary School and Presidential Awardee for Excellence in Science Teaching

Professional Societies
American Institute of Physics
Marc Brodsky, Executive Director
Stephanie Campbell, Desktop Publisher
Liz Dart Caron, Project Administrator
Jack Hehn, Director of Education
Jim Stith, Director, Physics Resources Center

American Physical Society
Judy Franz, Executive Officer
Darlene Logan, Development Manager
Fred Stein, Director of Education

American Association of Physics Teachers
Warren Hein, Associate Executive Officer
Bernie Khoury, Executive Officer